Contents

Get it right

Choose the correct answer.

1. The factors of 36 are
   - a) 6 and 8
   - b) 4 and 9
   - c) 4 and 8

2. The product of 8 and 9 is
   - a) 64
   - b) 56
   - c) 72

3. In $1\frac{1}{2}$ minutes there are
   - a) 30 seconds
   - b) 60 seconds
   - c) 90 seconds

4. An isosceles triangle has
   - a) 2 equal sides
   - b) 3 equal sides
   - c) no equal sides

5. 6·5 is the same as
   - a) $6\frac{1}{2}$
   - b) $6\frac{1}{4}$
   - c) 65

6. An equivalent fraction to $\frac{3}{8}$ is
   - a) $\frac{3}{4}$
   - b) $\frac{9}{12}$
   - c) $\frac{6}{16}$

7. If I make 72 ten times larger I will have
   - a) 7200
   - b) 7·2
   - c) 720

8. If I take 890 from 1000 my answer will be
   - a) more than 100
   - b) less than 100
   - c) more than 200

9. The average of 10, 15, 8 is
   - a) 11
   - b) 10
   - c) 15

10. In a square there are
    - a) 6 lines of symmetry
    - b) 4 lines of symmetry
    - c) 8 lines of symmetry

11. $3\frac{1}{4}$ l is less than
    - a) 3·200 l
    - b) 3·800 l
    - c) 3·050 l

12. $4\frac{1}{2}$ kg is larger than
    - a) 4·800 kg
    - b) 4·250 kg
    - c) 4·600 kg

13. The reflection of [flag] is:
    - a) 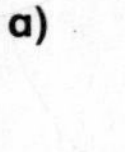
    - b) 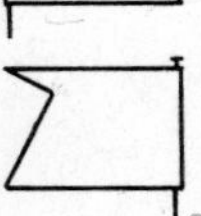
    - c) 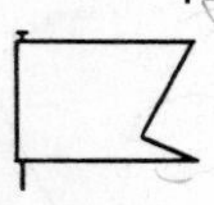

14. The area of the shape below is
    - a) $10\,cm^2$
    - b) $12\,cm^2$
    - c) $9\,cm^2$

3 cm
1 cm
4 cm
2 cm

15. A hexagon has
    a) 8 vertices
    b) 5 vertices
    c) 6 vertices

16. If I take 99 from 760 I have
    a) 859
    b) 661
    c) 659

17. $\frac{23}{12}$ is
    a) greater than 1
    b) less than 1
    c) greater than 2

18. $4\frac{5}{10}$ is the same as
    a) 4·5
    b) 5·4
    c) $4\frac{2}{5}$

19. If I make $\frac{1}{2}$ turn from facing NE I will face
    a) SE
    b) NW
    c) SW

20. If 29·1 is made 10 times bigger it will become
    a) 2910
    b) 2·9
    c) 291

21. $\frac{3}{5}$ of 100 is
    a) 150
    b) 60
    c) 90

22. From 0900 to 1530 is
    a) $8\frac{1}{2}$ hr
    b) 9 hr 30 min
    c) 6 hr 30 min

23. If I add $\frac{3}{4}$ and $\frac{5}{8}$ the answer will be
    a) more than 1
    b) less than 1
    c) more than 2

24. If 272 is made 10 times smaller the answer will be
    a) 2720
    b) 27·2
    c) 2072

25. 1000 mm is
    a) 1 m
    b) 10 m
    c) $\frac{1}{10}$ m

26. 24·5 is between
    a) 2 and 3
    b) 20 and 30
    c) 200 and 300

27. 85 × 10 is
    a) more than 1000
    b) less than 1000
    c) more than 2000

28. The circumference of a circle is
    a) the distance across it
    b) half the distance across it
    c) the distance round it

29. A square 100 cm × 100 cm has an area of
    a) 1 m^2
    b) 10 m^2
    c) 100 m^2

30. The volume of the shape below is
    a) 7 m^3
    b) 12 m^3
    c) 9 cm^3

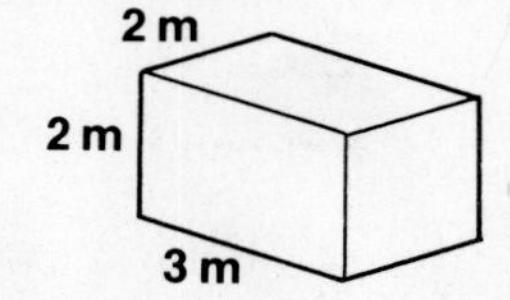

Number

Write the answers only

1. 274 + 9
2. 132 + 99
3. 513 + 99
4. 2314 + 999
5. 613 + 9
6. 4010 + 999
7. 2162 + 99
8. 6118 + 99
9. 2848 + 999
10. 4137 + 999

11.
```
  4293
    27
+  148
______

______
```

12.
```
  4314
  2998
+  175
______

______
```

13.
```
   728
  3904
+ 1669
______

______
```

14.
```
  5881
   373
+ 2965
______

______
```

15. Find the total of 64, 1287 and 3949.
16. Find the total value of £3·68, £2·99 and £0·95.
17. Increase £8·52 by £7·38.
18. Find an amount which is £7·50 greater than £12·75.

Write the answers only.

19. 194 − 9
20. 1348 − 99
21. 962 − 99
22. 735 − 9
23. 2416 − 999
24. 2311 − 999
25. 4062 − 99
26. 1003 − 999
27. 851 − 9
28. 483 − 99

29.
```
  6433
− 2487
______

______
```

30.
```
  5006
− 2984
______

______
```

31.
```
  1738
−  879
______

______
```

32.
```
  5900
− 3916
______

______
```

33. How much greater than 2483 is 5000?
34. How much do I need to make £3·75 into £10?
35. Decrease £9·53 by £1·85.
36. By how much is £8·50 less than £13·25?
37. How much less than £25 is £8·35?

Measurement

Remember:	0740 means 7.40 am
	1635 means 4.35 pm

These clocks show the time the 24 hour clock way.
Rewrite the times using am or pm.

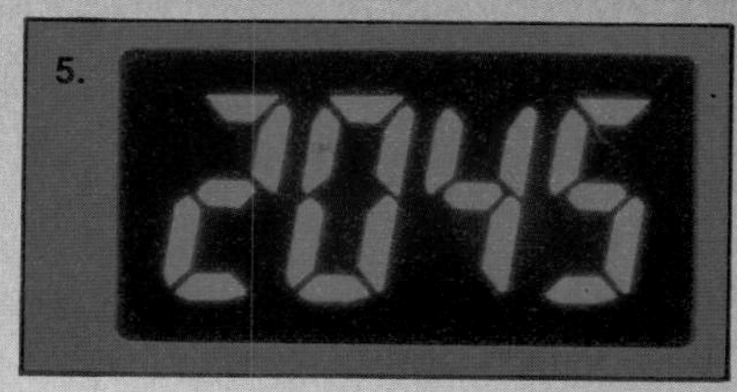

Write these times in the 24 hour clock way.

7. 4.30 pm
8. 11.45 am
9. 6.05 am
10. 12 noon

How many hours and minutes are there between these times?

11. 1245 and 1355
12. 0715 and 0950
13. 1630 and 1818

This table shows the times of sunrise and sunset on the first day of each month.

	sunrise	sunset		sunrise	sunset		sunrise	sunset
Jan.	0810	1610	May	0529	2031	Sept.	0612	1949
Feb.	0743	1655	June	0450	2110	Oct.	0700	1821
Mar.	0649	1746	July	0452	2118	Nov.	0710	1630
Apr.	0628	1942	Aug.	0530	2047	Dec.	0743	1556

14. How long from sunrise to sunset on January 1st?
15. How long from sunrise to sunset on May 1st?

Fractions

1. $\frac{3}{4} + \frac{2}{5}$
2. $\frac{1}{3} + \frac{3}{4}$
3. $\frac{1}{2} + \frac{6}{7}$
4. $\frac{3}{5} + \frac{2}{3}$
5. $\frac{5}{8} + \frac{2}{3} + \frac{1}{4}$
6. $\frac{1}{2} + \frac{2}{3} + \frac{1}{4}$
7. $\frac{1}{4} + \frac{3}{5} + \frac{1}{2}$
8. $\frac{1}{2} + \frac{5}{6} + \frac{2}{3}$
9. $\frac{3}{8} - \frac{1}{4}$
10. $\frac{9}{10} - \frac{1}{2}$
11. $\frac{3}{4} - \frac{1}{3}$
12. $\frac{1}{2} - \frac{1}{3}$
13. $\frac{7}{8} - \frac{1}{6}$
14. $\frac{5}{6} - \frac{3}{4}$
15. $\frac{3}{4} - \frac{2}{5}$
16. $\frac{1}{2} - \frac{2}{7}$

17. What fraction of the rectangle is covered by each shape?
18. What fraction of the rectangle is covered by shapes C + E?
19. What fraction of the rectangle is covered by shapes A + B + C?

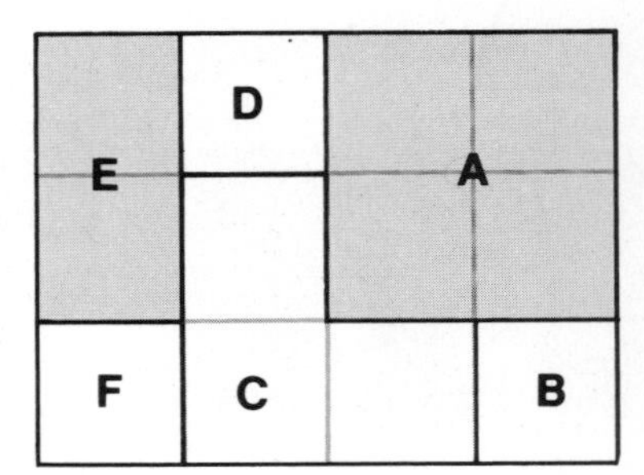

Remember: The **denominator** tells you the number of parts into which the whole one has been divided. The **numerator** tells you how many of those parts there are.

$\frac{3}{4}$ ← numerator / denominator

Fill in the missing numerators and denominators.

20. $\frac{1}{2} = \frac{*}{4} = \frac{*}{6} = \frac{*}{8} = \frac{5}{*} = \frac{6}{*} = \frac{*}{16}$
21. $\frac{1}{4} = \frac{*}{8} = \frac{*}{12} = \frac{4}{*} = \frac{5}{*} = \frac{*}{24}$
22. $\frac{3}{4} = \frac{*}{8} = \frac{9}{*} = \frac{*}{16} = \frac{15}{*} = \frac{18}{*}$
23. $\frac{1}{3} = \frac{*}{6} = \frac{3}{*} = \frac{4}{*} = \frac{*}{18} = \frac{*}{24}$
24. $\frac{2}{3} = \frac{*}{6} = \frac{6}{*} = \frac{8}{*} = \frac{*}{18} = \frac{*}{24}$

When making fractions equivalent, the numerator and denominator are multiplied by the same number.

$$\frac{3 \times \mathbf{4}}{4 \times \mathbf{4}} = \frac{12}{16} \qquad \frac{2 \times \mathbf{3}}{3 \times \mathbf{3}} = \frac{6}{9}$$

Fill in the missing numerators and denominators.

1. $\frac{10}{20} = \frac{5}{*} = \frac{*}{2}$
2. $\frac{4}{12} = \frac{2}{*} = \frac{1}{*}$
3. $\frac{6}{18} = \frac{*}{6} = \frac{1}{*}$
4. $\frac{6}{12} = \frac{*}{4} = \frac{*}{2}$
5. $\frac{18}{24} = \frac{9}{*} = \frac{*}{4}$
6. $\frac{4}{16} = \frac{*}{8} = \frac{*}{4}$

When making fractions equivalent, the numerator and denominator are divided by the same number.

$$\frac{8 \div \mathbf{4}}{12 \div \mathbf{4}} = \frac{2}{3} \qquad \frac{6 \div \mathbf{2}}{10 \div \mathbf{2}} = \frac{3}{5}$$

$\frac{12}{16}$ can be **reduced** to $\frac{3}{4}$

When a fraction cannot be reduced any further, it is in its simplest form.

$\frac{12}{16}$ in its **simplest form** is $\frac{3}{4}$

Reduce each of these to its simplest form.

7. $\frac{2}{6} = \frac{*}{3}$
8. $\frac{4}{12} = \frac{*}{3}$
9. $\frac{10}{15} = \frac{*}{3}$
10. $\frac{15}{18} = \frac{*}{6}$
11. $\frac{15}{24} = \frac{*}{8}$
12. $\frac{14}{21} = \frac{2}{*}$
13. $\frac{15}{21} = \frac{*}{7}$
14. $\frac{16}{20} = \frac{4}{*}$
15. $\frac{6}{12}$
16. $\frac{15}{20}$
17. $\frac{14}{16}$
18. $\frac{20}{30}$
19. $\frac{12}{16}$
20. $\frac{15}{25}$
21. $\frac{10}{12}$
22. $\frac{12}{15}$
23. $\frac{18}{24}$
24. $\frac{10}{16}$
25. $\frac{16}{20}$
26. $\frac{9}{12}$

Reduce each answer to its simplest form.

27. $\frac{1}{2} + \frac{5}{6}$
28. $\frac{5}{6} + \frac{2}{3}$
29. $\frac{7}{8} + \frac{7}{8}$
30. $\frac{3}{4} + \frac{11}{12}$
31. $\frac{1}{6} + \frac{1}{2}$
32. $\frac{1}{4} + \frac{5}{12}$
33. $\frac{5}{8} + \frac{7}{8}$
34. $\frac{3}{4} + \frac{5}{12}$
35. $\frac{1}{10} + \frac{1}{2}$
36. $\frac{1}{3} + \frac{5}{12}$
37. $\frac{1}{2} + \frac{9}{10}$
38. $\frac{7}{10} + \frac{3}{10}$

Shape

A regular pentagon

A regular hexagon

A regular octagon

A **regular** shape has all its sides and angles equal.

Copy and complete this table.

Shape	Number of equal sides	Number of equal angles	Number of right angles	Number of lines of symmetry	Number of vertices
Square					
Rectangle					
Scalene triangle					
Isosceles triangle					
Equilateral triangle					
Regular hexagon					

A B C D E F G H I J K L M N O P Q R S T U V W X Y Z

Some capital letters have lines of symmetry.

Write the capital letters which have

1. one line of symmetry
2. two lines of symmetry
3. more than two lines of symmetry
4. no lines of symmetry

Protractor, compasses

Use a protractor and ruler to draw these triangles.

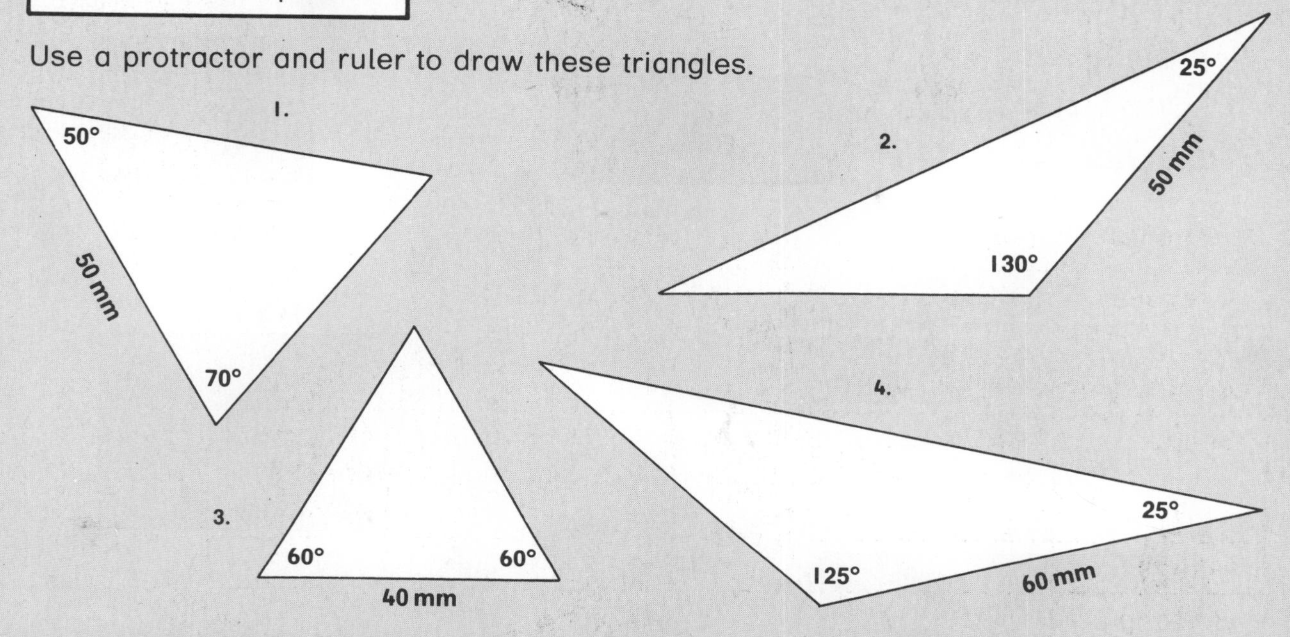

Measure the third angle.
Measure the other two sides.
Record the measurements of each triangle.
Write the name of each triangle.

Draw these shapes.

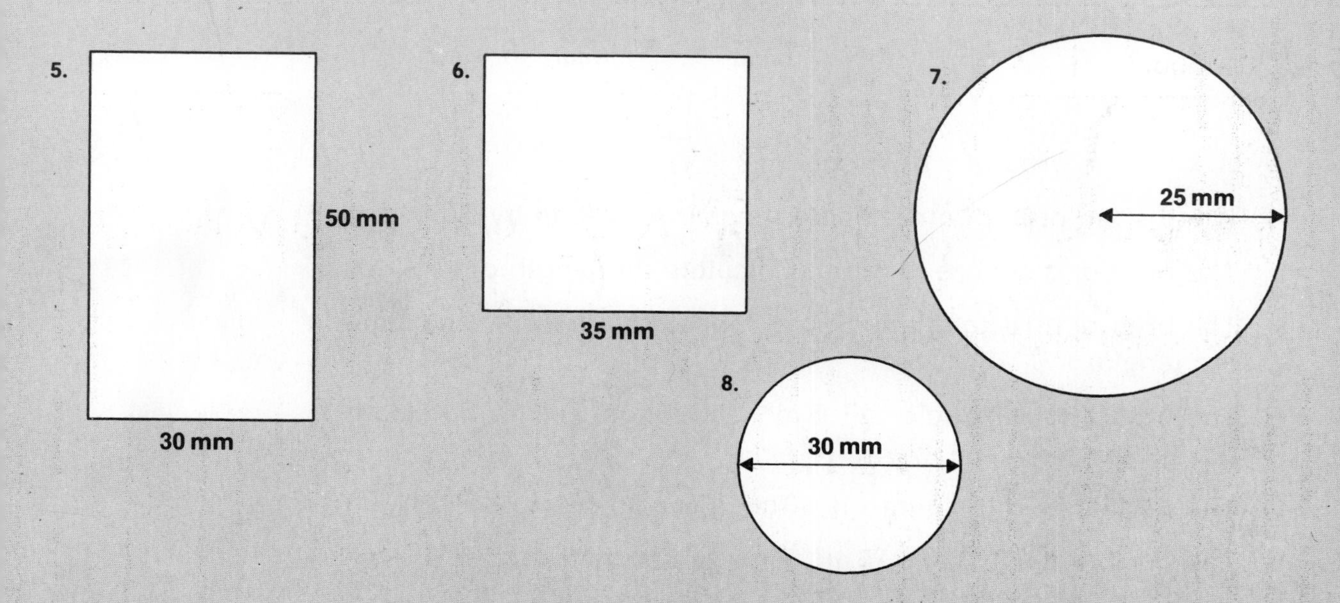

Area

Calculate these areas.

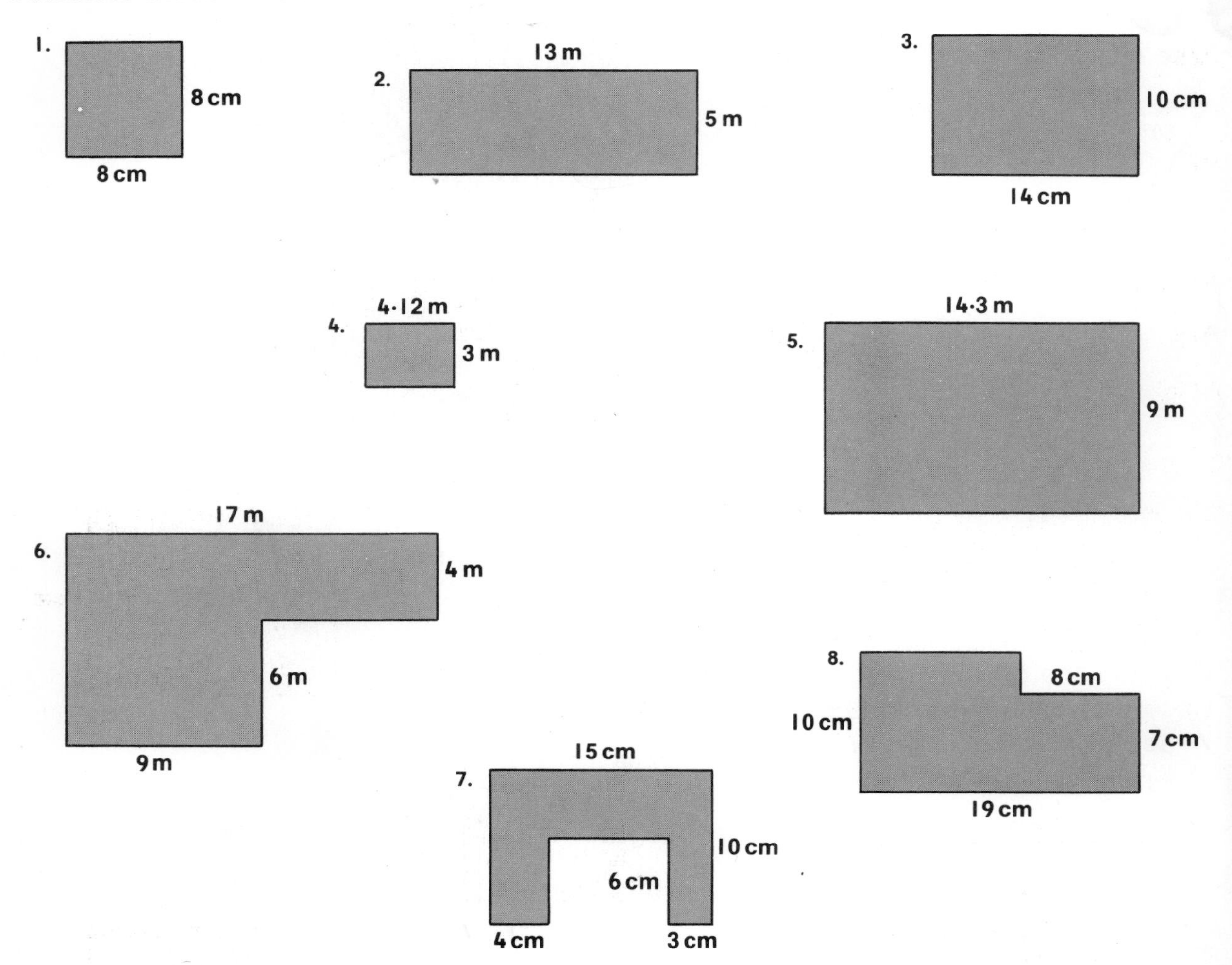

9. What is the area of a 6 cm square?
10. What is the area of a rectangle measuring 3·42 m by 5 m?
11. The area of a square is 81 m^2. Calculate the length of its sides.
12. The area of a rectangle is 54 cm^2. The width is 9 cm. Calculate the length.
13. The area of a rectangle is 108 m^2. The length is 9 m. Calculate the width.
14. The perimeter of a square is 40 cm. Find its area.
15. The area of a square is 25 m^2. Find its perimeter.

Calculate the area of the shaded part.

1\. (10 cm by 8 cm rectangle; unshaded square 2 cm)

Area of whole shape = ☐ cm^2

Area of unshaded part = ☐ cm^2

Area of shaded part = ☐ cm^2

2\. (12 cm by 6 cm rectangle; unshaded rectangle 4 cm by 2 cm)

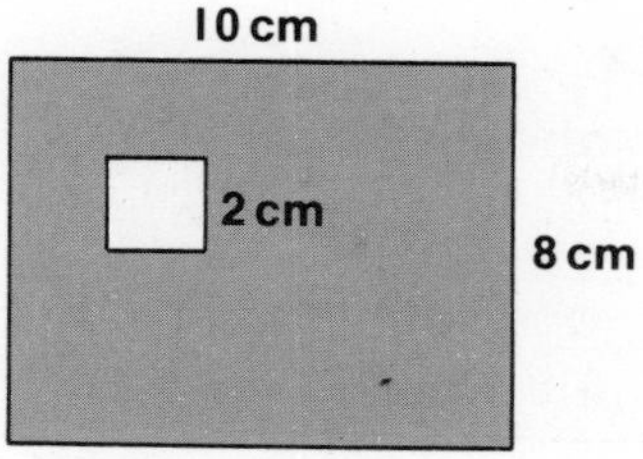

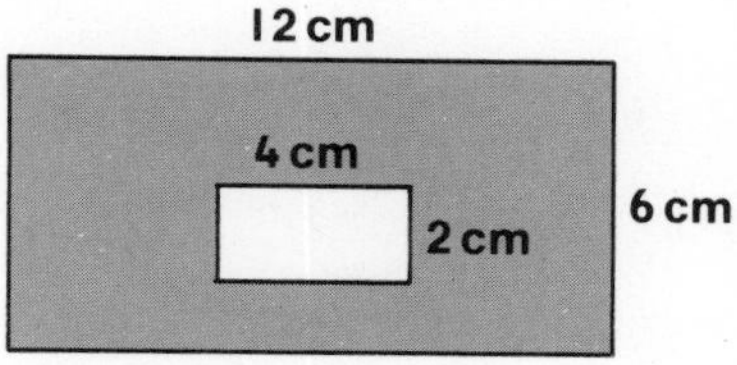

Area of whole shape = ☐ cm^2

Area of unshaded part = ☐ cm^2

Area of shaded part = ☐ cm^2

3\. 15 cm, 6 cm; 12 cm, 2 cm

4\. 10 cm, 9 cm

5\. 14 cm, 8 cm; 4 cm

6\. 8 cm, 8 cm

7\. 8 cm, 7 cm

8\. 10 cm, 8 cm; 4 cm

9\. 8 cm, 8 cm; 3 cm

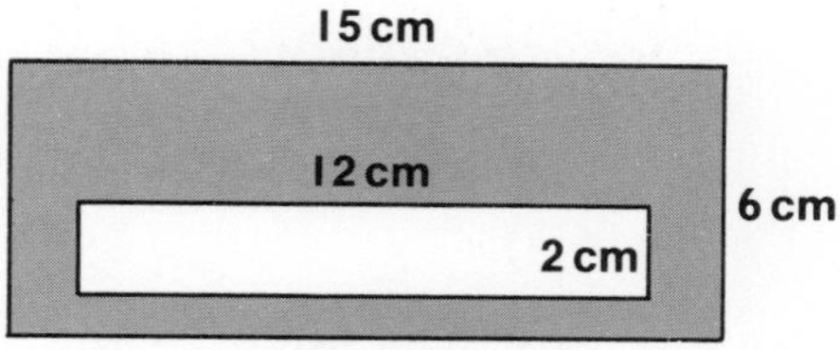

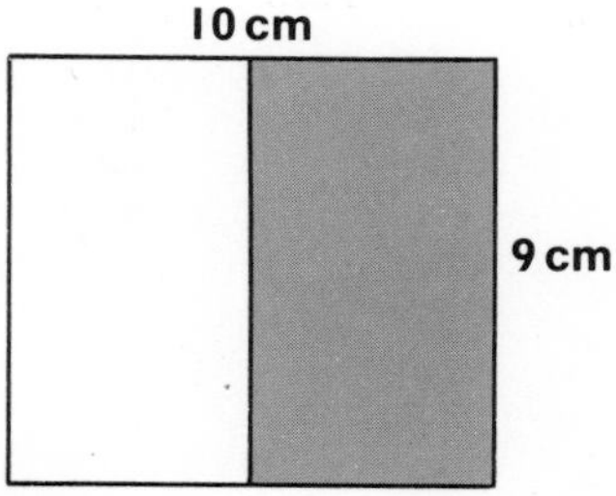

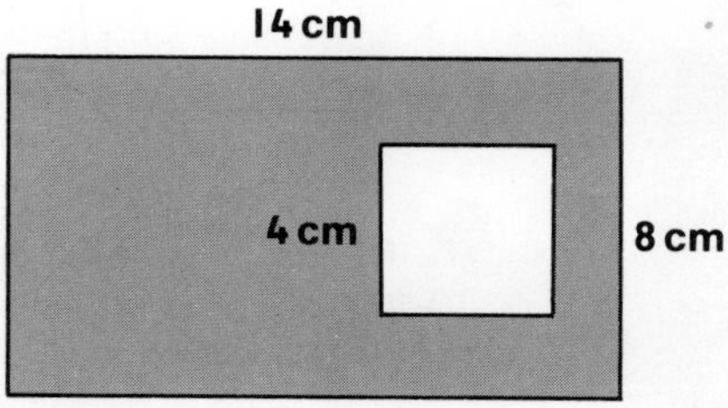

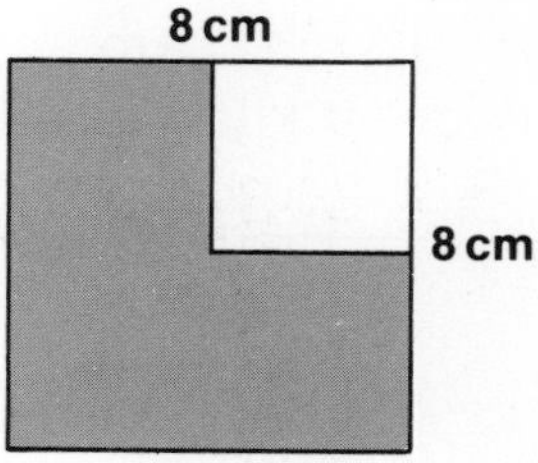

10\. 9 cm, 4 cm

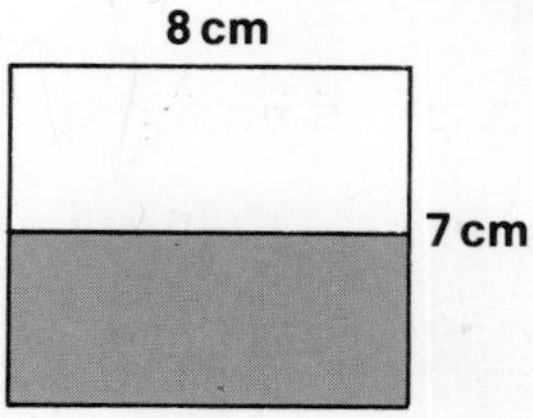

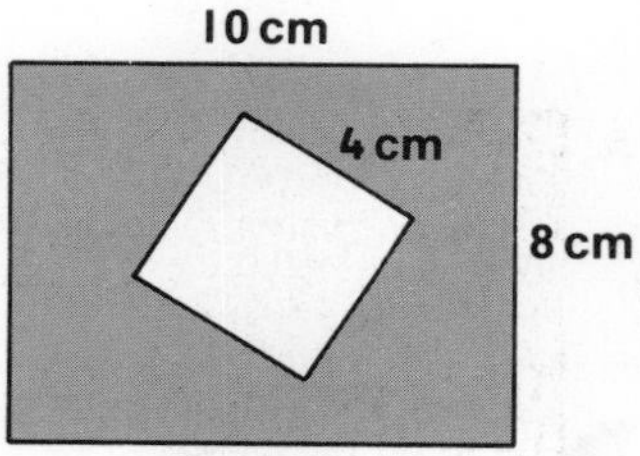

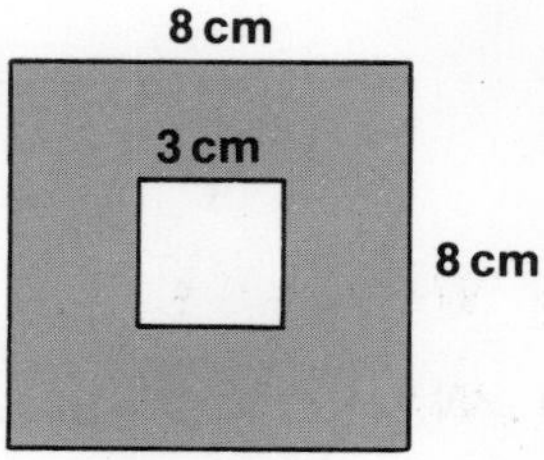

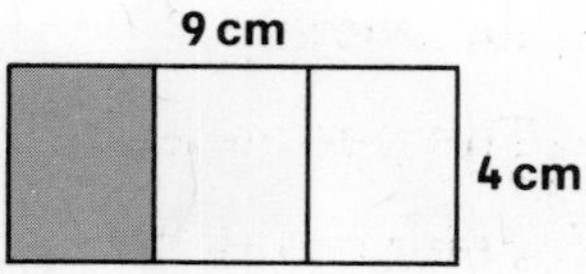

Graphs

Cm² paper

Draw a horizontal axis labelled from 0 to 14, and a vertical axis labelled from 0 to 16.

Plot these co-ordinates, and join them up in the order given.
(3.1) (11.1) (14.4) (0.4) (3.1)

On the same axes plot these co-ordinates and join them up.
(0.5) (8.5) (8.16) (0.5)
(9.5) (13.5) (9.14) (9.5)

What is the picture you have drawn?

Draw another pair of axes.
Label the horizontal axis from 0 to 14.
Label the vertical axis from 0 to 8.

Plot these co-ordinates, and join them up in the order given.
(0.0) (2.1) (2.3) (4.3) (5.4) (7.7) (8.5) (11.8) (11.6) (14.4)

You have plotted the journey taken by the vessel you drew earlier.

The journey is drawn to a scale of 1 cm : 1 km.
Measure and calculate to the nearest kilometre how far the vessel travelled.

This column graph shows the time taken by seven yachts to complete a race.

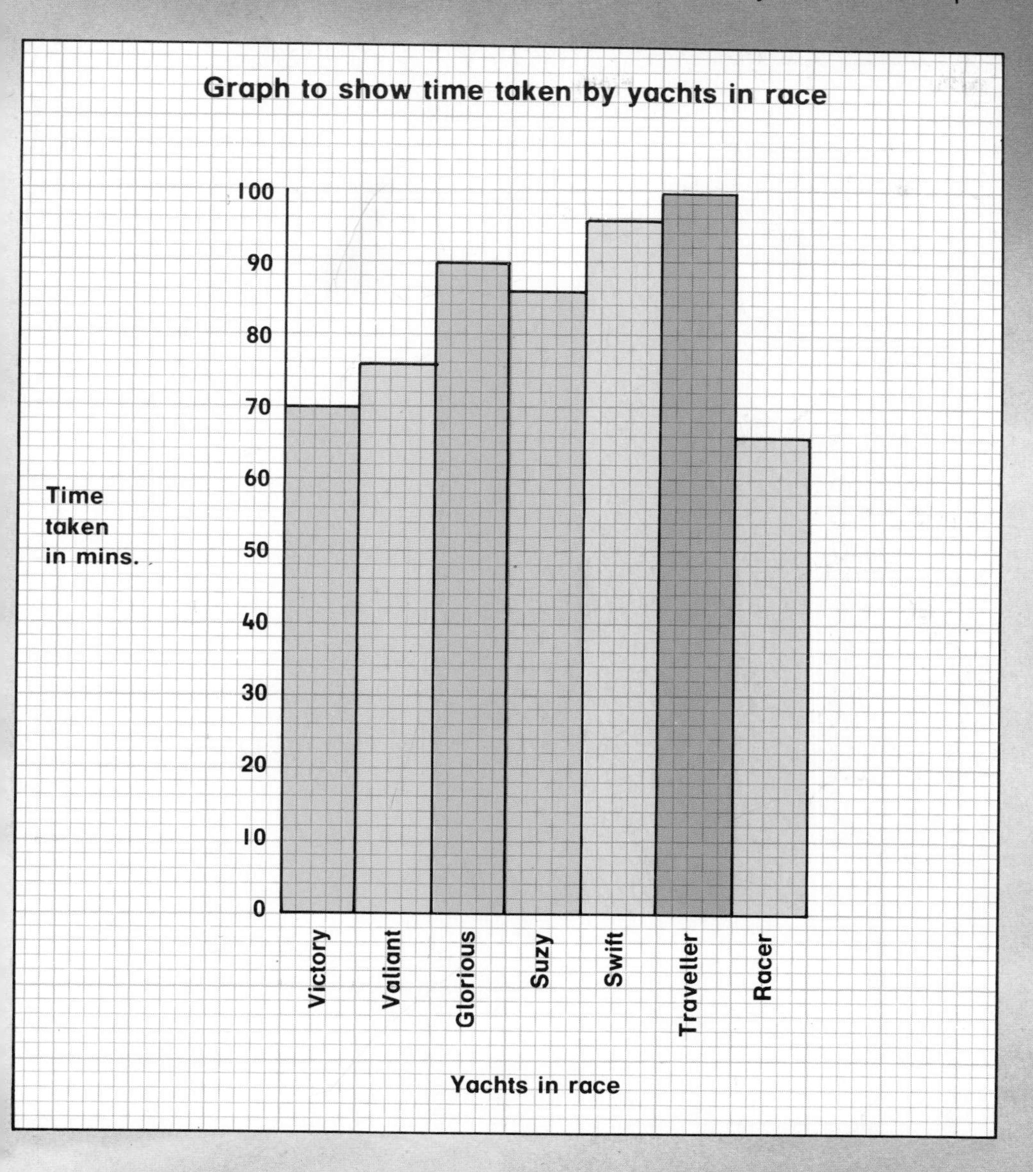

1. Which yacht won the race?
2. Write the order in which the yachts finished the race.
3. Write the time taken by each yacht.
4. What was the difference in time between the winner and the loser?

Fractions

When subtracting mixed numbers, take away the whole numbers first.

$$3\frac{3}{4} - 1\frac{3}{8}$$
$$= 2\frac{6}{8} - \frac{3}{8}$$
$$= 2\frac{3}{8}$$

1. $4\frac{1}{2} - 3\frac{1}{3}$
2. $3\frac{5}{8} - 2\frac{1}{3}$
3. $5\frac{11}{12} - 1\frac{3}{4}$
4. $2\frac{7}{8} - 2\frac{1}{5}$
5. $4\frac{6}{7} - \frac{1}{2}$
6. $3\frac{9}{10} - 2\frac{2}{3}$
7. $3\frac{4}{5} - 1\frac{1}{2}$
8. $2\frac{2}{3} - 2\frac{1}{4}$
9. $1\frac{3}{4} - \frac{2}{5}$
10. $2\frac{2}{5} - 1\frac{1}{6}$
11. $3\frac{5}{9} - 2\frac{1}{6}$
12. $4\frac{2}{3} - 1\frac{2}{7}$

When you cannot subtract the fraction, decompose a whole one.

$$4\frac{1}{3} - 2\frac{1}{2}$$
$$= 2\frac{2}{6} - \frac{3}{6}$$
$$= 1\frac{8}{6} - \frac{3}{6}$$
$$= 1\frac{5}{6}$$

13. $5\frac{1}{5} - 1\frac{2}{3}$
14. $2\frac{1}{4} - 1\frac{5}{6}$
15. $4\frac{3}{8} - 2\frac{1}{2}$
16. $6\frac{1}{3} - \frac{3}{4}$
17. $4\frac{3}{10} - 3\frac{3}{4}$
18. $3\frac{1}{6} - 1\frac{1}{4}$
19. $3\frac{1}{2} - 2\frac{5}{7}$
20. $5\frac{2}{5} - 4\frac{3}{4}$
21. $2\frac{1}{3} - \frac{1}{2}$
22. $4\frac{5}{12} - 1\frac{7}{8}$
23. $3\frac{2}{3} - 1\frac{3}{4}$
24. $5\frac{2}{9} - 2\frac{5}{6}$
25. $4\frac{1}{2} - 2\frac{9}{10}$
26. $6\frac{2}{3} - 1\frac{3}{4}$
27. $3\frac{1}{6} - 2\frac{3}{4}$
28. $5\frac{1}{3} - 2\frac{3}{5}$
29. $4\frac{1}{3} - 1\frac{5}{7}$
30. $7\frac{4}{9} - \frac{5}{6}$
31. $5\frac{1}{4} - 4\frac{5}{7}$
32. $2\frac{1}{10} - \frac{1}{4}$

1. $2\frac{5}{7} + 3\frac{2}{3}$
2. $1\frac{5}{6} + 2\frac{1}{2}$
3. $2\frac{5}{6} + 1\frac{3}{4}$
4. $3\frac{1}{2} + \frac{3}{5}$
5. $\frac{3}{4} + 2\frac{5}{6}$
6. $2\frac{1}{4} + 2\frac{6}{7}$
7. $\frac{2}{3} + 1\frac{7}{8}$
8. $3\frac{3}{5} + 1\frac{2}{3}$
9. $2\frac{9}{10} + 1\frac{1}{2}$
10. $3\frac{5}{8} + 4\frac{3}{4}$
11. $5\frac{2}{3} + 2\frac{4}{5}$
12. $1\frac{7}{10} + 3\frac{5}{6}$
13. $1\frac{3}{4} + 5\frac{2}{3}$
14. $4\frac{6}{7} + 3\frac{2}{3}$
15. $2\frac{5}{6} + 1\frac{3}{4}$
16. $5\frac{1}{3} + 3\frac{1}{2}$
17. Find the total of $4\frac{1}{2}$ and $2\frac{4}{7}$.
18. Add together $3\frac{5}{6}$ and $2\frac{7}{8}$.
19. Find the sum of $5\frac{7}{8}$ and $1\frac{2}{3}$.
20. $3\frac{3}{4}$ plus $1\frac{2}{3}$.
21. Add $2\frac{2}{3}$ to $3\frac{4}{5}$.
22. Find the missing fraction: $* - 1\frac{3}{4} = 3\frac{2}{5}$.
23. Find the total of $6\frac{7}{8}$ and $1\frac{5}{6}$.
24. $1\frac{1}{3} - \frac{1}{2}$
25. $3\frac{5}{12} - 2\frac{5}{8}$
26. $3 - \frac{2}{3}$
27. $4\frac{1}{8} - 3\frac{5}{6}$
28. $5\frac{1}{6} - 2\frac{3}{5}$
29. $3\frac{2}{5} - 1\frac{3}{4}$
30. $6 - \frac{7}{8}$
31. $2\frac{2}{9} - 1\frac{1}{2}$
32. $4\frac{5}{12} - 2\frac{2}{3}$
33. $4\frac{2}{9} - 1\frac{3}{4}$
34. $4\frac{3}{5} - 2\frac{7}{8}$
35. $3\frac{3}{10} - 1\frac{3}{4}$
36. $8 - 4\frac{3}{5}$
37. $4\frac{1}{5} - 2\frac{3}{4}$
38. $6\frac{3}{10} - 2\frac{3}{4}$
39. $9 - 4\frac{3}{8}$
40. Subtract $4\frac{6}{7}$ from $6\frac{1}{3}$.
41. $3\frac{1}{6}$ minus $1\frac{3}{4}$.
42. Find the difference between $1\frac{1}{4}$ and $3\frac{1}{5}$.
43. From $2\frac{1}{2}$ take $1\frac{2}{3}$.
44. How much less than $3\frac{1}{4}$ is $1\frac{7}{8}$?
45. How much more is $4\frac{2}{7}$ than $1\frac{4}{5}$?
46. Take $2\frac{3}{4}$ from $3\frac{1}{7}$.

Number

Write the answers only.

1. 262 × 10
2. 138 × 10
3. 471 × 10
4. £1·80 × 10
5. £2·72 × 10
6. £3·24 × 10
7. £10·10 × 10
8. £7·50 × 10

9. $\begin{array}{r} 136 \\ \times \quad 9 \\ \hline \\ \hline \end{array}$
10. $\begin{array}{r} 207 \\ \times \quad 8 \\ \hline \\ \hline \end{array}$
11. $\begin{array}{r} 1352 \\ \times \quad 6 \\ \hline \\ \hline \end{array}$
12. $\begin{array}{r} 1098 \\ \times \quad 7 \\ \hline \\ \hline \end{array}$
13. $\begin{array}{r} 1850 \\ \times \quad 4 \\ \hline \\ \hline \end{array}$

14. 418 × 8
15. 1359 × 5
16. 2139 × 4
17. 4036 × 2
18. £2·78 × 7
19. £11·39 × 3
20. £7·81 × 9
21. £0·86 × 7
22. Multiply three thousand two hundred by 3.
23. Make 1528 five times greater.
24. What amount is 6 times as much as £3·75?
25. What amount is double £29·75?
26. Find the product of 9 and 807.
27. Find a number which is 7 times as large as 884.

Write the answers only.

28. 2460 ÷ 10
29. 3720 ÷ 10
30. 5160 ÷ 10
31. 2000 ÷ 10
32. 6090 ÷ 10
33. £5·50 ÷ 10
34. £10·90 ÷ 10
35. £5·40 ÷ 10
36. $3\overline{)3723}$
37. $5\overline{)4720}$
38. $6\overline{)6369}$
39. $9\overline{)1428}$
40. £7·60 ÷ 8
41. £12·62 ÷ 2
42. £10·57 ÷ 7
43. £13·68 ÷ 8
44. Divide 7482 by 9.
45. How many 8s are there in 2048?
46. Share £15·75 equally among 7 people.
47. Find the average of £1·50, £2·80 and £1·10.
48. Find $\frac{2}{3}$ of £23·10.
49. Find $\frac{3}{8}$ of £15·60.

This hotel has bedrooms on five floors.

This table shows how many bedrooms there are on each floor and the number of beds in each room.

Floor	No. of rooms with		
	1 bed	2 beds	3 beds
1st	5	6	5
2nd	3	2	5
3rd	8	2	3
4th	6	4	4
5th	5	4	2

1. How many bedrooms are there on each floor?
2. How many bedrooms does the hotel have altogether?
3. How many beds are there on each floor?
4. How many beds does the hotel have?
5. If $\frac{3}{4}$ of the beds on the 1st floor are booked, how many are empty on that floor?
6. If $\frac{1}{3}$ of the beds in the hotel are empty, how many people are sleeping there?

Graphs

This graph shows the weight of potatoes a man dug from his garden over a period of 8 years.

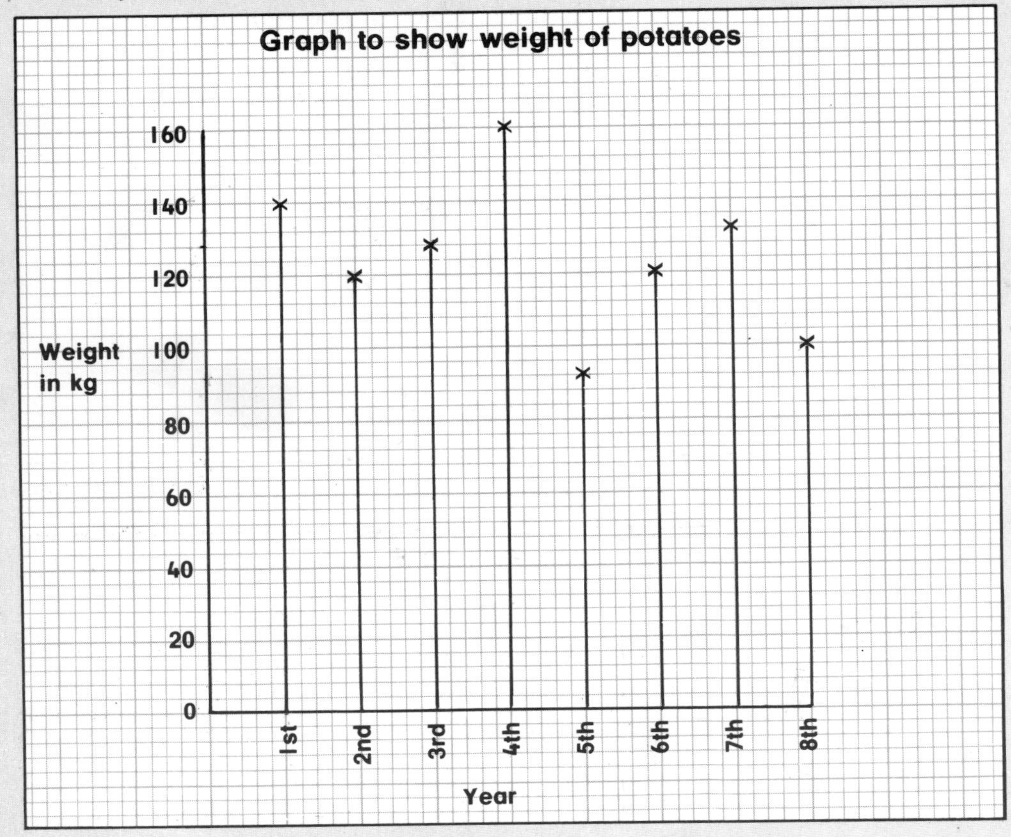

1. How many kg of potatoes did he grow each year?
2. What was the total weight of potatoes he grew in the 8 years?
3. He sold $\frac{1}{4}$ of his crop each year. How many kg did he sell each year?
4. This table shows the price per kg he charged each year.

Year	1st	2nd	3rd	4th	5th	6th	7th	8th
Price per kg	6p	7p	9p	8p	12p	17p	13p	14p

   What was his income from potatoes each year?

5. What was his total income from potatoes over the 8 years?

Shape

Compasses

Use a ruler and compasses to construct this triangle.

Step 1

Draw a line 50 mm long.
Label the line AB.

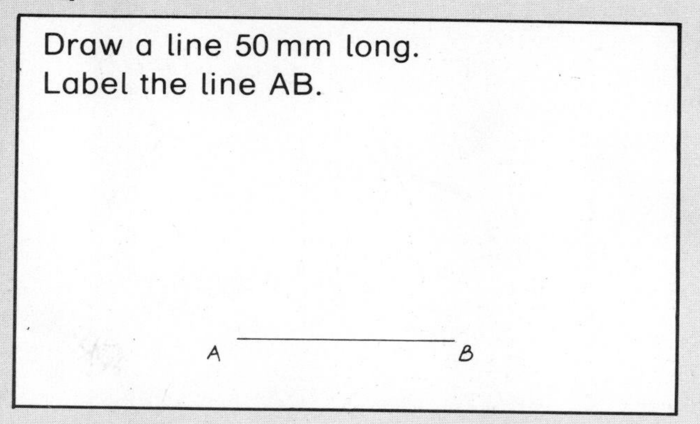

Step 2

Set your compasses at 50 mm.
With compass point on A, draw an arc.

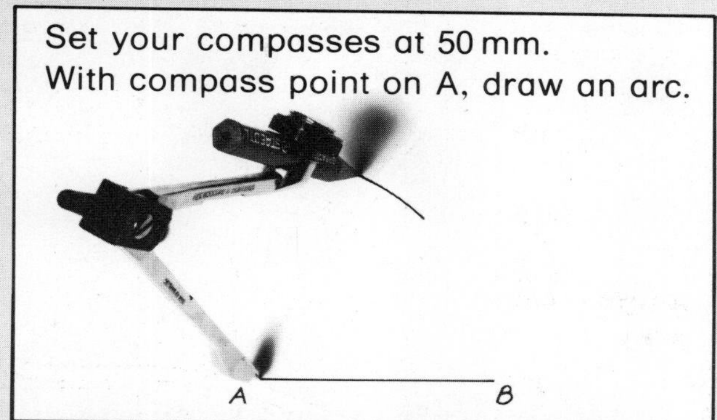

Step 3

With compasses still set at 50 mm, now put the point on B. Draw an arc to intersect at C.

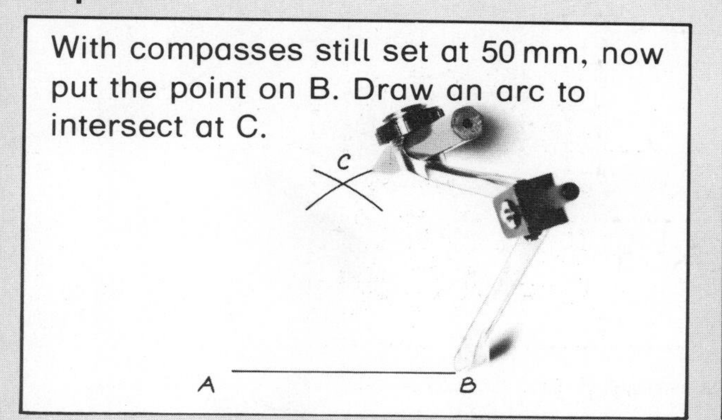

Step 4

Join AC and BC.

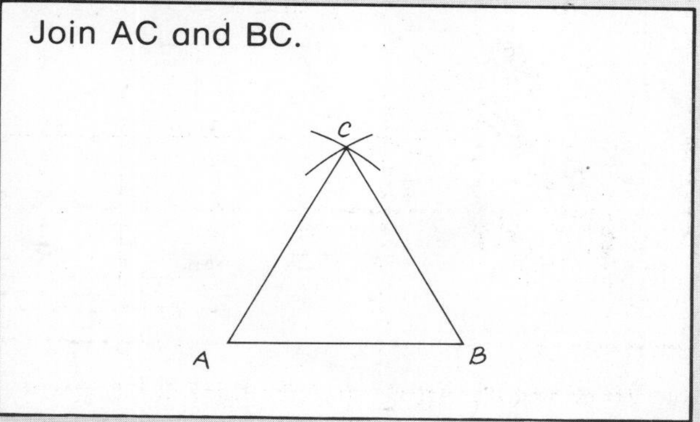

Write the name of the triangle you have constructed.

Construct an equilateral triangle whose sides measure 35 mm.

Draw a circle of radius 30 mm.
With compasses still set at 30 mm, mark off arcs round the circumference.
Join adjacent intersections with straight lines.
Write the name of the shape you have constructed.

Construct a hexagon in a circle whose radius is 45 mm.

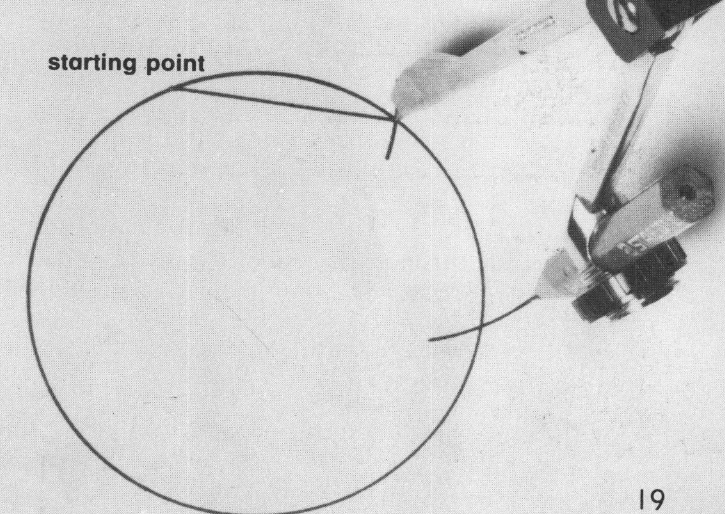

Measurement

This map shows the hamlet of Copley.

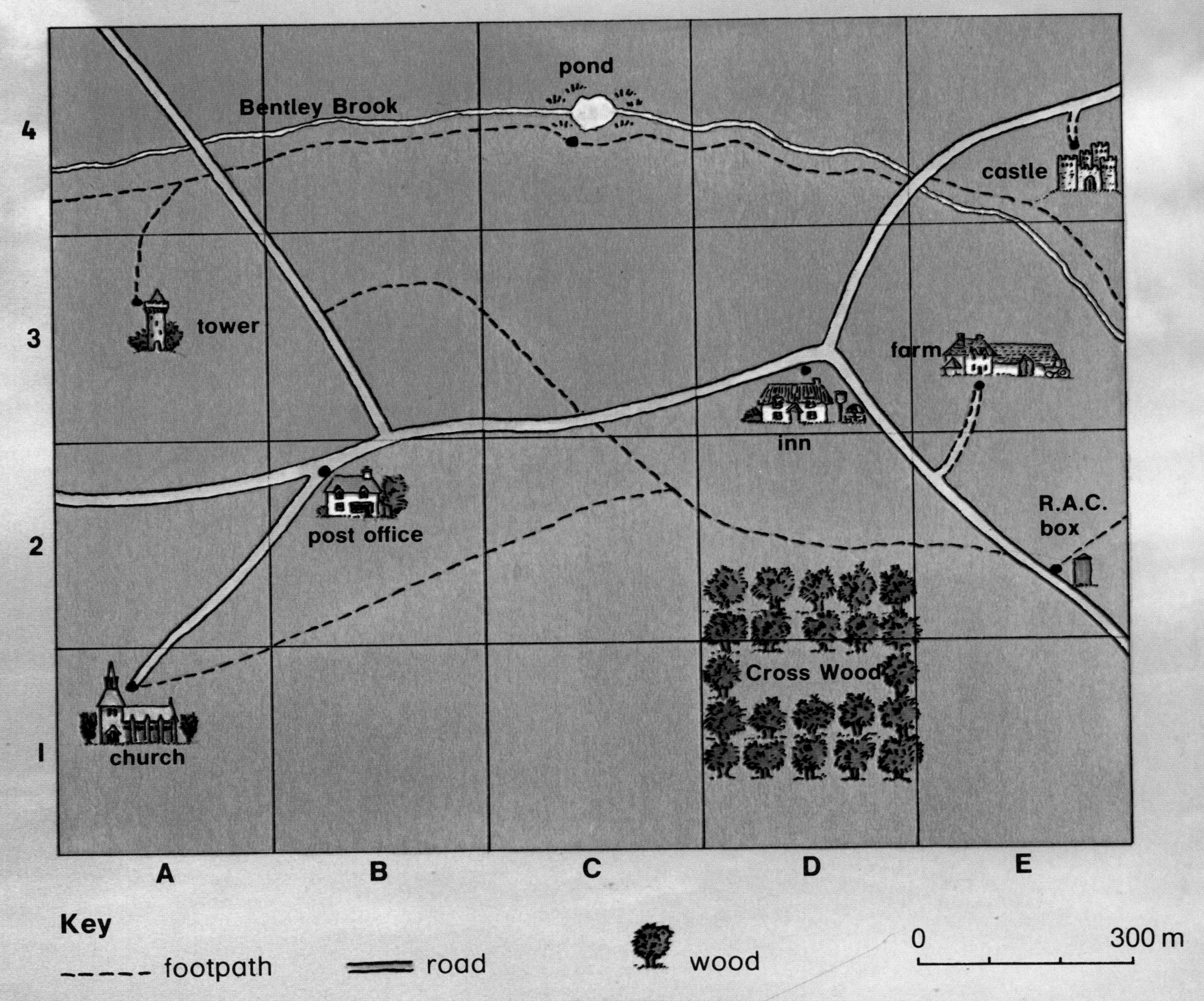

Using roads and footpaths find the shortest distances between these points.

1. castle : inn
2. R.A.C. box : church
3. post office : pond
4. tower : farm
5. R.A.C. box : pond
6. post office : farm

Which buildings will you find in the following squares?

7. A1
8. D3
9. E3
10. A3
11. B2

12. Calculate the perimeter of Cross Wood.

Remember: $>$ means greater than
$<$ means less than
$=$ means equal to

The signs are missing.
Copy and put in the correct sign.

1. $3\frac{1}{2}$ m * 320 cm
2. $5\frac{1}{2}$ cm * 52 mm
3. $3\frac{1}{4}$ kg * 3·200 kg
4. $1\frac{3}{4}$ l * 1850 ml
5. 2·400 kg * $2\frac{1}{2}$ kg
6. 1600 g * 1·600 kg
7. $1\frac{3}{4}$ m * 1·75 m
8. $2\frac{1}{2}$ l * 2·600 l
9. $3\frac{1}{2}$ cm * 38 mm
10. 4·140 l * 4410 ml
11. 3450 g * 3·500 kg
12. $4\frac{1}{2}$ km * 4500 m

13. 3 mm $\xrightarrow{\times 10}$ $\square$ cm
14. 2 m $\xrightarrow{\div 10}$ $\square$ cm
15. 2 kg $\xrightarrow{\div 10}$ $\square$ g
16. 20 ml $\xrightarrow{\times 100}$ $\square$ l
17. 35 m $\xrightarrow{\times 100}$ $\square$ km
18. 60 cm $\xrightarrow{\div 100}$ $\square$ mm
19. 50 g $\xrightarrow{\times 100}$ $\square$ kg

Which is the greater?

20. $\frac{1}{2}$ of 3 kg or $\frac{1}{4}$ of 5 kg
21. $\frac{3}{4}$ of 1 l or $\frac{2}{3}$ of 900 ml
22. $\frac{1}{2}$ of $1\frac{1}{2}$ m or $\frac{1}{4}$ of 1 m
23. $\frac{3}{4}$ of 2 kg or $\frac{1}{2}$ of 3 kg

24. 1·5 kg = $\square$ g
25. 2·5 l = $\square$ ml
26. 0·5 m = $\square$ cm
27. 0·2 kg = $\square$ g
28. 1·2 l = $\square$ ml
29. 3·5 m = $\square$ cm
30. 2·7 kg = $\square$ g
31. 0·7 l = $\square$ ml
32. 1·7 m = $\square$ cm

Number

Do these as quickly as you can.

1. 6 × 8 2. 4 × 9 3. 3 × 10 4. 7 × 7 5. 8 × 6

6. 9 × 8 7. 5 × 10 8. 7 × 8 9. 9 × 9 10. 4 × 5

11. 2 × 9 12. 10 × 10 13. 6 × 7 14. 4 × 8 15. 9 × 7

16. 8 × 8 17. 3 × 9 18. 10 × 6 19. 8 × 3 20. 8 × 7

21. 5 × 5 22. 5 × 8 23. 10 × 7 24. 4 × 6 25. 6 × 6

26. 7 × 9 27. 2 × 7 28. 4 × 10 29. 8 × 9 30. 7 × 6

31. 264 × 7 32. 408 × 9 33. 1202 × 8 34. 705 × 6 35. 1403 × 5

36. 1318 × 4 37. 874 × 10 38. 1169 × 7 39. 3169 × 3 40. 849 × 8

Write the factors of these numbers.

41. 14 42. 16 43. 24 44. 18 45. 32 46. 20 47. 36

Factors can be used for multiplication.
To multiply by 8, we can multiply by its factors.

$17 \times 8 \quad = \quad 17 \times 4 \times 2 \quad = \quad 17 \times 2 \times 4$

$$\begin{array}{r} 17 \\ \times \quad 8 \\ \hline 136 \\ \hline \end{array} \qquad \begin{array}{r} 17 \\ \times \quad 4 \\ \hline 68 \\ \times \quad 2 \\ \hline 136 \\ \hline \end{array} \qquad \begin{array}{r} 17 \\ \times \quad 2 \\ \hline 34 \\ \times \quad 4 \\ \hline 136 \\ \hline \end{array}$$

We get the same answer each way.
We call this multiplication by factors.

Multiply the following numbers by factors.
Decide your factors first before you multiply.

48. 129 × 8 49. 207 × 12 50. 188 × 12 51. 193 × 15 52. 124 × 16

53. 403 × 20 54. 145 × 18 55. 108 × 21 56. 136 × 30 57. 107 × 14

58. 416 × 9 59. 169 × 24 60. 148 × 15 61. 192 × 25 62. 151 × 28

63. 532 × 20 64. 98 × 80 65. 142 × 50 66. 111 × 70 67. 76 × 90

Use the quick method to multiply each of these by 10.

1. 36
2. 84
3. 116
4. 229
5. 328
6. 147
7. 209
8. 390
9. 612
10. 300
11. 844
12. 632

To multiply by 20, you can multiply by factors.
Multiply by 10 first, then by 2.

$216 \times 20 = 2160 \times 2 = 4320$

Multiply each of these by 20.

13. 18
14. 46
15. 213
16. 327
17. 105
18. 293
19. 478
20. 72
21. 418
22. 490
23. 372
24. 340

Now do these in the same way.

25. 36×30
26. 116×40
27. 132×50
28. 37×60
29. 421×20
30. 236×40
31. 113×80
32. 64×90
33. 108×70
34. 323×30

35. Copy and complete this table.

	× 20	× 30	× 40	× 50	× 60	× 70	× 80	× 90
112								
231								

Choose the correct answer from the box.

563
632
2000
540
93
4150
50
9460
3060

36. Multiply 27×20.
37. Find $\frac{1}{10}$ of 5630.
38. Find a number which is 20 times greater than 473.
39. What must 106 be multiplied by to give 5300?
40. Make 102 thirty times larger.
41. What is the product of 50 and 83?
42. How many times will 10 go into 6320?
43. Add (106×10) to (94×10).
44. Which number when multiplied by 20 gives an answer of 1860?

Measurement

Various cylinders, bow caliper, sliding caliper

Measure the diameter of each cylinder.
Here are some ways of doing it.

Put the cylinder between two blocks of wood and measure their distance apart.

Use bow calipers.

Use a sliding caliper.

Measure the circumference of each cylinder.
Here are some ways of doing it.

Put a mark on the circumference of the cylinder.
Roll the cylinder for one turn.
Mark how far it has rolled.
Measure to find the circumference.

Measure with string.

Record all your results in a table.

cylinder	diameter	circumference
1		
2		
3		

Approximately how many times will the diameter divide into the circumference?

Fractions

1. Draw a line $\frac{3}{4}$ the length of this line.

2. Draw a rectangle whose perimeter is $\frac{4}{5}$ as long as this one.

3. Draw a circle whose radius is $\frac{2}{3}$ as long as this one.

4. Draw a square whose sides are $\frac{3}{4}$ the length of this square.

5. Draw a square whose area is $\frac{1}{4}$ of the size of this square.

6. Draw a shape whose area is half the size of this shape.

7. Draw a circle whose diameter is $\frac{1}{2}$ as long as the one below.

Decimals

Make each of these numbers 10 times smaller.

1. 4200	2. 324	3. 6247	4. 907	5. 1002
6. 5060	7. 24	8. 8249	9. 8	10. 7000

Make each of these numbers 10 times bigger.

11. 14·8	12. 240	13. 170·9	14. 39·7	15. 900
16. 429·1	17. 60·2	18. 5	19. 123·4	20. 72
21. 18·6	22. 158·1	23. 96	24. 421·6	25. 18·9

Make each of these numbers 100 times smaller.

26. 3000	27. 410	28. 6290	29. 40	30. 730
31. 4200	32. 8120	33. 290	34. 90	35. 7020
36. 180	37. 7240	38. 80	39. 4160	40. 9820

Make each of these numbers 100 times bigger.

41. 4·9	42. 27	43. 13·7	44. 0·8	45. 32·1
46. 92	47. 80·6	48. 1·2	49. 5·6	50. 89·7
51. 3·7	52. 71·1	53. 92	54. 54·4	55. 68

70 made ten times smaller is 7

7 made ten times smaller is $\frac{7}{10}$

$\frac{7}{10}$ made ten times smaller is $\frac{7}{100}$

$\frac{7}{100}$ made ten times smaller is $\frac{7}{1000}$

T	U		$\frac{1}{10}$	$\frac{1}{100}$	$\frac{1}{1000}$
7	0	·			
	7	·			
	0	·	7		
	0	·	0	7	
	0	·	0	0	7

0·7 can be written as $\frac{7}{10}$

0·07 can be written as $\frac{7}{100}$

0·007 can be written as $\frac{7}{1000}$

Write the value of the underlined digit.

1. 0·0$\underline{8}$	**2.** 0·0$\underline{4}$	**3.** 0·1$\underline{7}$	**4.** 0·$\underline{5}$2	**5.** 0·6$\underline{9}$
6. 0·00$\underline{4}$	**7.** 0·00$\underline{9}$	**8.** 0·02$\underline{4}$	**9.** 0·09$\underline{2}$	**10.** 0·1$\underline{3}$7
11. 0·$\underline{4}$23	**12.** 0·50$\underline{2}$	**13.** 6·1$\underline{2}$4	**14.** $\underline{3}$·847	**15.** 2·07$\underline{8}$

Make each of these numbers 10 times smaller.

16. 0·17	**17.** 1·06	**18.** 2·4	**19.** 37	**20.** 0·02
21. 3·4	**22.** 1·09	**23.** 5·22	**24.** 17·04	**25.** 0·99

Make each of these numbers 10 times bigger.

26. 0·746	**27.** 1·478	**28.** 0·072	**29.** 1·4	**30.** 0·009
31. 16	**32.** 14·7	**33.** 2·078	**34.** 1·001	**35.** 0·108

Make each of these numbers 100 times smaller.

36. 527	**37.** 0·4	**38.** 1212	**39.** 20	**40.** 24·7
41. 4617·8	**42.** 4200	**43.** 208	**44.** 607·9	**45.** 710·7

Make each of these numbers 100 times bigger.

46. 8·32	**47.** 0·149	**48.** 72·06	**49.** 24·006	**50.** 1·428
51. 17·84	**52.** 15·692	**53.** 0·002	**54.** 0·017	**55.** 4·044

Make each of these numbers 1000 times smaller.

56. 2000	**57.** 79	**58.** 5208	**59.** 120	**60.** 6
61. 3614	**62.** 17	**63.** 10	**64.** 408	**65.** 392

Make each of these numbers 1000 times bigger.

66. 0·009	**67.** 6	**68.** 2·784	**69.** 7·24	**70.** 0·17
71. 0·7	**72.** 0·058	**73.** 0·092	**74.** 1·098	**75.** 0·101

Number

Do these a quick way.

1. 126×10
2. 234×10
3. 118×20
4. 137×20
5. 87×30
6. 212×40
7. 181×50
8. 98×20
9. 64×90
10. 137×60

172×12 can be written as $(172 \times 10) + (172 \times 2)$.

We can work it out like this:

$$\begin{array}{r} 172 \\ \times\ 12 \\ \hline 1720 \\ +\ 344 \\ \hline 2064 \\ \hline \end{array} \quad \begin{array}{l} \\ \\ \\ \Leftarrow 172 \times 10 \\ \Leftarrow 172 \times 2 \\ \Leftarrow 172 \times 12 \end{array}$$

128×24 can be written as $(128 \times 20) + (128 \times 4)$.

We can work it out like this:

$$\begin{array}{r} 128 \\ \times\ 24 \\ \hline 2560 \\ +\ 512 \\ \hline 3072 \\ \hline \end{array} \quad \begin{array}{l} \\ \\ \\ \Leftarrow 128 \times 20 \\ \Leftarrow 128 \times 4 \\ \Leftarrow 128 \times 24 \end{array}$$

Work these out.

11. 114×13
12. 218×16
13. 164×21
14. 203×17
15. 168×22
16. 149×18
17. 65×31
18. 183×15
19. 34×16
20. 403×24

We can multiply money in the same way.

$$\begin{array}{r} \text{£} \\ 2\cdot24 \\ \times \quad 16 \\ \hline 22\cdot40 \\ +\ 13\cdot44 \\ \hline 35\cdot84 \\ \hline \end{array} \qquad \begin{array}{l} \\ \\ \\ \\ \Leftarrow \text{£}2\cdot24 \times 10 \\ \Leftarrow \text{£}2\cdot24 \times 6 \\ \Leftarrow \text{£}2\cdot24 \times 16 \end{array}$$

$$\begin{array}{r} \text{£} \\ 1\cdot76 \\ \times \quad 23 \\ \hline 35\cdot20 \\ +\ 5\cdot28 \\ \hline 40\cdot48 \\ \hline \end{array} \qquad \begin{array}{l} \\ \\ \\ \\ \Leftarrow \text{£}1\cdot76 \times 20 \\ \Leftarrow \text{£}1\cdot76 \times 3 \\ \Leftarrow \text{£}1\cdot76 \times 23 \end{array}$$

Now do these:

21. £1·32 × 12
22. £1·48 × 15
23. £1·64 × 21
24. £2·13 × 17
25. £3·04 × 11
26. £5·16 × 14
27. £3·13 × 24
28. £2·82 × 13
29. £1·75 × 22
30. £1·38 × 19
31. £0·59 × 16
32. £2·44 × 25

From the table:

33. Find out how many people each cinema holds.

34. Which cinema holds most people?

	No. of rows	No. of seats per row
RITZY	27	35
GRANADA	24	39
RIALTO	25	38
CLASSIC	36	28
ODEON	23	32
ABC	28	31

Decimals

1. $\begin{array}{r} 2\cdot784 \\ 3\cdot904 \\ +\ 6\cdot078 \\ \hline \end{array}$

2. $\begin{array}{r} 15\cdot6 \\ 26\cdot07 \\ +\ \ 1\cdot418 \\ \hline \end{array}$

3. $\begin{array}{r} 0\cdot124 \\ 0\cdot7 \\ +\ 0\cdot857 \\ \hline \end{array}$

4. $\begin{array}{r} 0\cdot076 \\ 2\cdot42 \\ +\ 9\cdot987 \\ \hline \end{array}$

5. $\begin{array}{r} 2\cdot63 \\ 0\cdot9 \\ +\ 3\cdot076 \\ \hline \end{array}$

6. $\begin{array}{r} 4\cdot507 \\ 8\cdot8 \\ +\ 0\cdot598 \\ \hline \end{array}$

7. $\begin{array}{r} 5\cdot724 \\ 7\cdot581 \\ +\ 3\cdot795 \\ \hline \end{array}$

8. $\begin{array}{r} 14\cdot24 \\ 30\cdot5 \\ +\ \ 0\cdot485 \\ \hline \end{array}$

9. Add 16·742, 0·074 and 1·849.
10. Find the total of 0·7, 3·48 and 0·019.
11. 1·72 plus 2·076.
12. Add together 17, 13·4 and 16·05.

13. $\begin{array}{r} 8\cdot076 \\ -\ 2\cdot438 \\ \hline \end{array}$

14. $\begin{array}{r} 17\cdot421 \\ -\ \ 9\cdot834 \\ \hline \end{array}$

15. $\begin{array}{r} 0\cdot72 \\ -\ 0\cdot684 \\ \hline \end{array}$

16. $\begin{array}{r} 7\cdot49 \\ -\ 2\cdot583 \\ \hline \end{array}$

17. $\begin{array}{r} 2\cdot4 \\ -\ 0\cdot146 \\ \hline \end{array}$

18. $\begin{array}{r} 12\cdot402 \\ -\ \ 3\cdot084 \\ \hline \end{array}$

19. $\begin{array}{r} 6\cdot005 \\ -\ 2\cdot998 \\ \hline \end{array}$

20. $\begin{array}{r} 13\cdot4 \\ -\ \ 0\cdot768 \\ \hline \end{array}$

21. Find the difference between 0·684 and 1·72.
22. 12·702 − 1·009
23. Take 1·49 from 9·321.
24. Subtract 17·47 from 20.
25. How much greater than 1·32 is 6·4?
26. From 7·315 take 2·9.
27. 5·36 − 2·984
28. How much more than 7·9 is 14·65?

1. $3{\cdot}693 \times 8$

2. $4{\cdot}076 \times 4$

3. $5{\cdot}71 \times 9$

4. $17{\cdot}08 \times 5$

5. $0{\cdot}147 \times 7$

6. $0{\cdot}68 \times 3$

7. $13{\cdot}8 \times 6$

8. $0{\cdot}079 \times 9$

9. $2{\cdot}308 \times 5$

10. $0{\cdot}523 \times 8$

11. $13{\cdot}016 \times 4$

12. $16{\cdot}29 \times 7$

13. Multiply 4·938 by 8.

14. 0·437 × 7

15. Find the product of 5 and 17·46.

16. 9 × 1·068

17. Multiply 16·026 by 9.

18. Find the product of 3 and 12·049.

19. $9\overline{)5{\cdot}211}$

20. $7\overline{)6{\cdot}041}$

21. $5\overline{)4{\cdot}525}$

22. $3\overline{)4{\cdot}734}$

23. $8\overline{)19{\cdot}84}$

24. $6\overline{)11{\cdot}016}$

25. $4\overline{)14{\cdot}36}$

26. $9\overline{)21{\cdot}636}$

27. $8\overline{)40{\cdot}896}$

28. $7\overline{)44{\cdot}73}$

29. $5\overline{)1305{\cdot}5}$

30. $8\overline{)12{\cdot}64}$

31. Divide 11·04 by 4.

32. 15·66 ÷ 9

33. 5·236 ÷ 7

34. Divide 6·444 by 6.

35. 2·360 ÷ 8

36. Divide 19·025 by 5.

Measurement

Protractor

Jason has to look up to see the top of the tree.
The angle he looks up is called the **angle of elevation**.

Measure these angles of elevation.

1.

2.

3.

4.

Clinometer, trundle wheel

You can find the approximate heights of tall objects using a clinometer.
Here is how Jason and Sally found the height of a tree.

They stood 10 metres from the tree.
They measured the distance with a trundle wheel.
They found the angle of elevation with a clinometer. It was 30°.

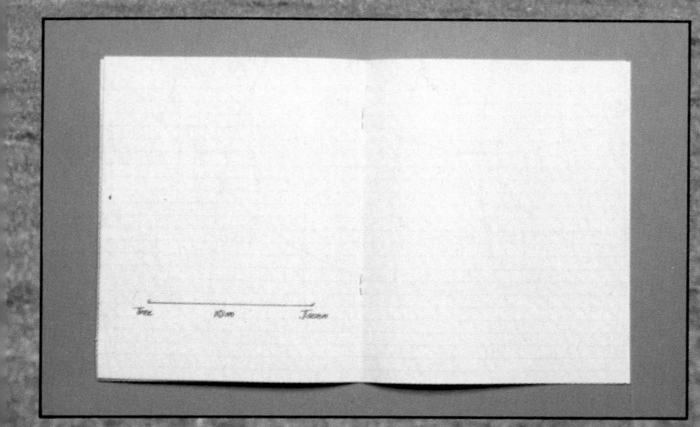

Back in the classroom they chose a scale of 1 cm : 1 m for their drawing.
They drew a line to show their distance from the tree.

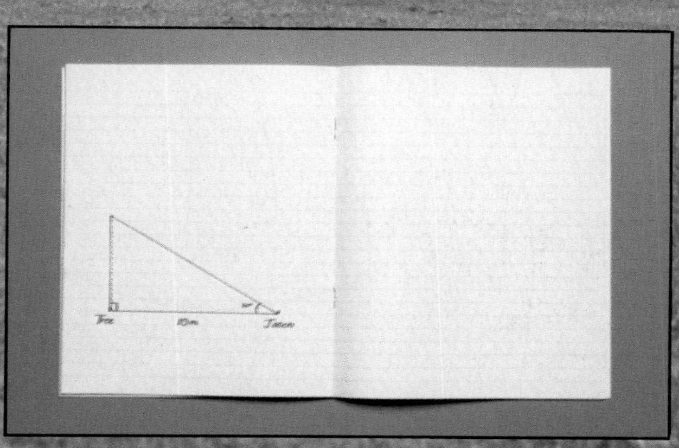

They drew the angle of elevation.
They then found the height of the tree.
Ask your teacher how they did it.

Measure the heights of three tall trees or buildings using a clinometer.
You may have to use a different scale from Sally and Jason.

Shape

Compasses

Remember: ⪢ This means copy into your book.

Draw a circle with a radius of 30 mm.
Set the compasses at 10 mm and mark off round the circumference. (The 10 mm marks will not fit exactly round the circumference.)
Label the top mark A.
Using each intersection as a centre, draw circles.
Each circle must pass through A.

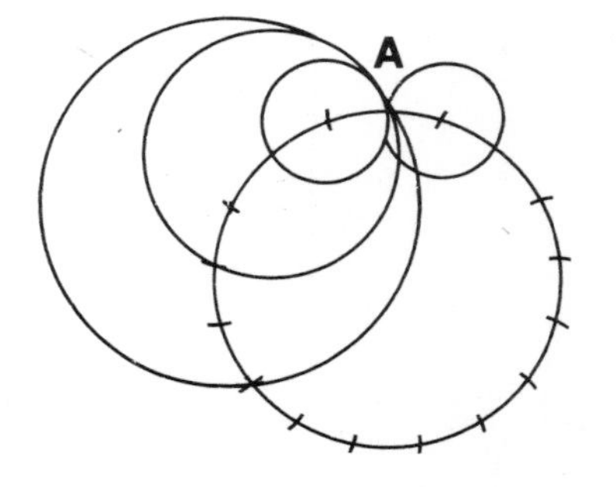

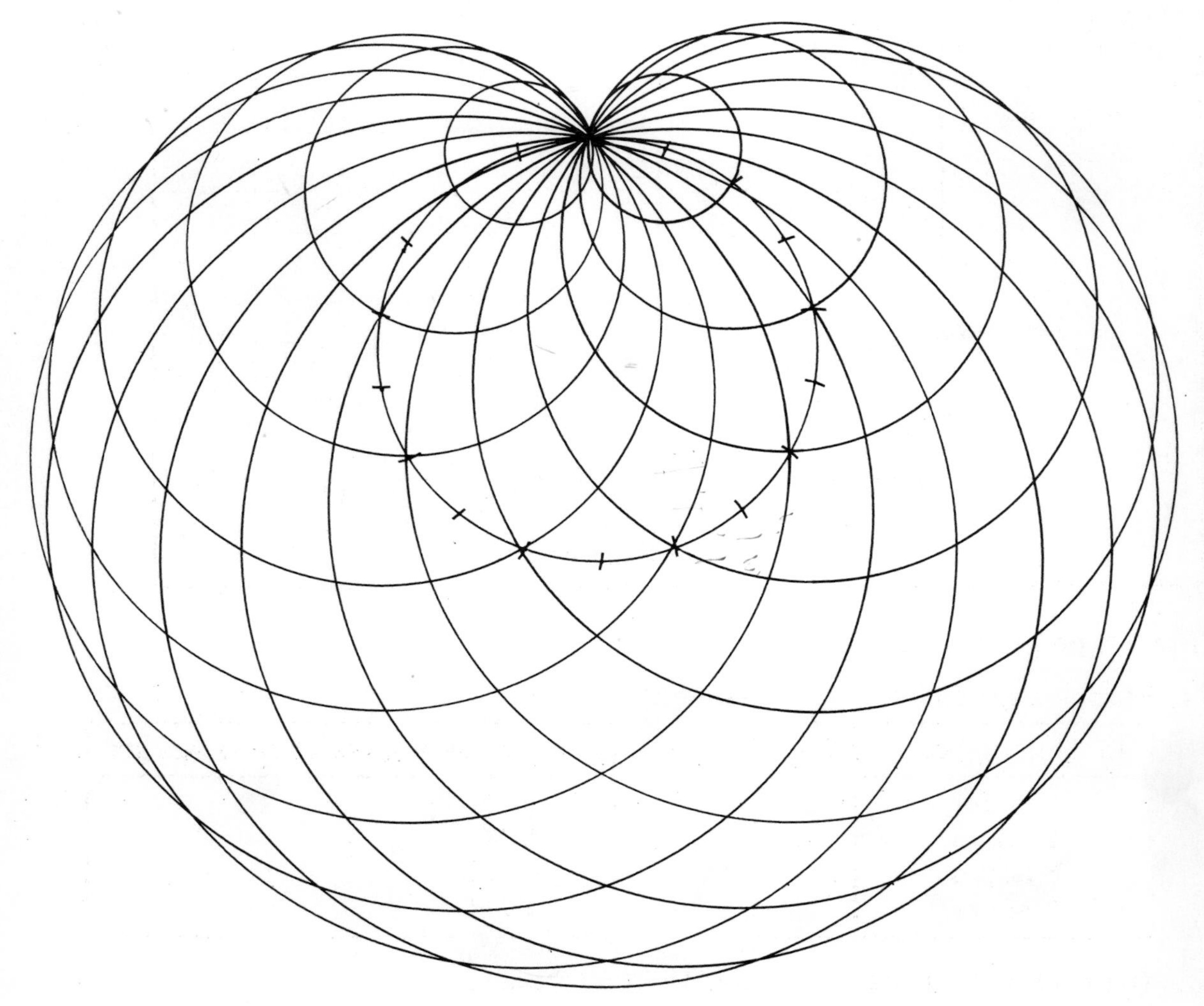

⪢ This pattern is called a **cardioid**.

Graphs

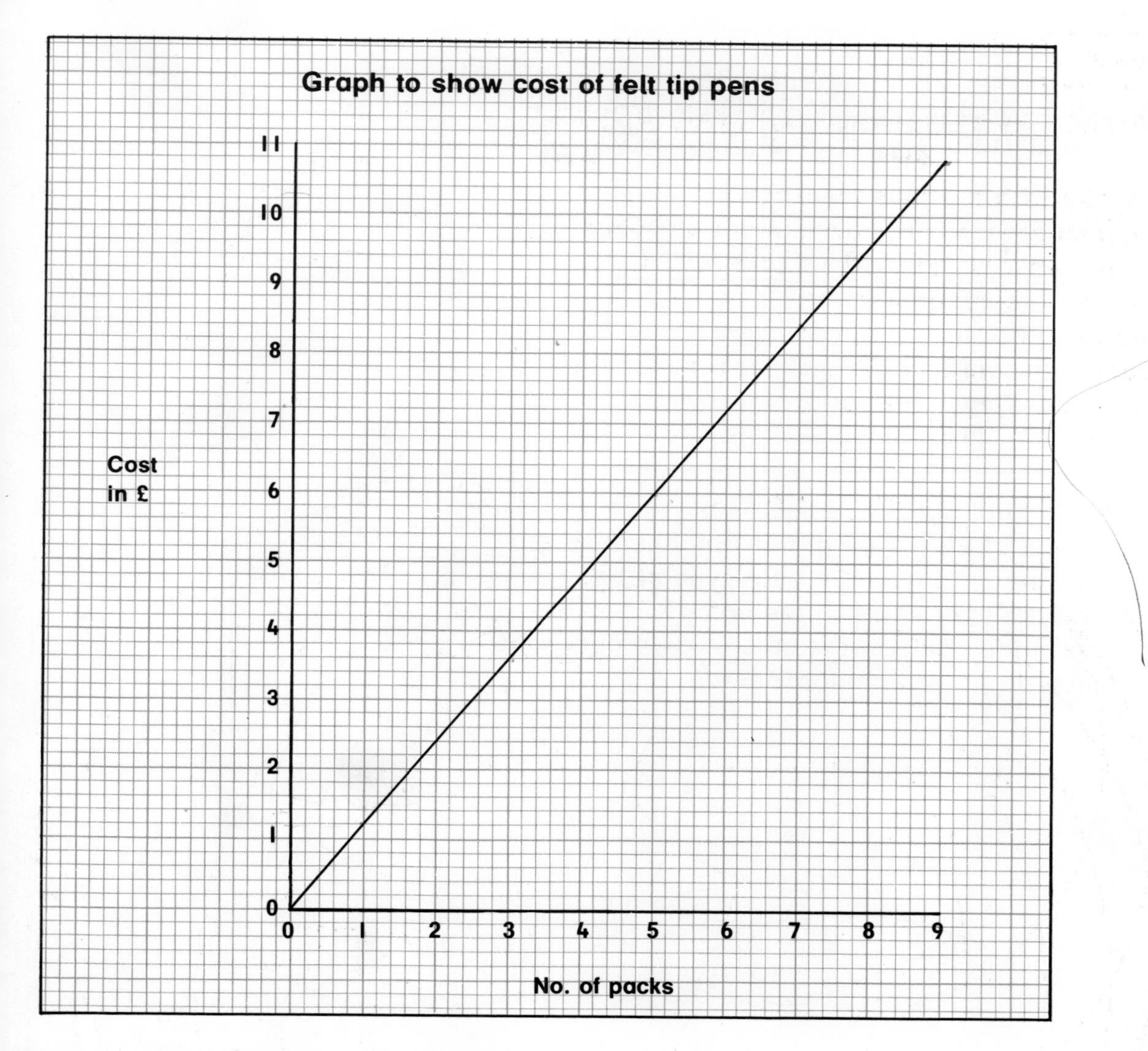

The felt tip pens were sold in packs of 12.

1. From the graph complete these tables.

No. of pens bought	Cost
1 pack	
2 packs	
5 packs	
7 packs	

Money spent	No. of pens bought
£1·20	
£3·60	
£10·80	
£7·20	

2. The cost of the pens was raised to £1·50 a pack.
Draw a line graph to show their cost at the new price.
You may need new scales on the vertical axis.

Number

Many trawler boats fish in the North Sea.
The fish they catch are put in big baskets to be sold.
This table shows how much fish one trawler caught and the value of the catch.

Fish	No. of baskets	Price per basket
Cod	147	£18
Plaice	116	£23
Halibut	109	£17
Hake	77	£13
Herring	82	£19

1. Each basket holds 12 kg of fish.
   What weight of each type of fish has been caught?
2. What was the total value of each type of fish caught?
3. What was the total value of the catch?
4. Exactly $\frac{2}{3}$ of the value of the catch was profit.
   How much profit was made from the trip?

This is a cross-channel ferry.
It takes vehicles and passengers across the Channel.
It makes one return journey each day.

These tables show the ferry charges and the number of vehicles and passengers the ferry carried one day last summer.

Charges (single fares)	
Cars and vans	£12
Coaches	£28
Lorries	£32
Passengers	£6

Outward journey	
Cars and vans	76
Coaches	13
Lorries	11
Passengers	258

Return journey	
Cars and vans	83
Coaches	12
Lorries	15
Passengers	234

1. How many vehicles were carried during the day?
2. How many passengers were carried during the day?
3. Calculate the income from each journey.
4. How much was earned by the ferry that day?

Measurement

1. $$\begin{array}{r} \text{kg} \\ 3\cdot478 \\ 2\cdot364 \\ +\ 1\cdot577 \\ \hline \\ \hline \end{array}$$

2. $$\begin{array}{r} \text{l} \\ 0\cdot350 \\ 2\cdot443 \\ +\ 1\cdot784 \\ \hline \\ \hline \end{array}$$

3. $$\begin{array}{r} \text{kg} \\ 3\cdot200 \\ -\ 1\cdot635 \\ \hline \\ \hline \end{array}$$

4. $$\begin{array}{r} \text{m} \\ 7\cdot50 \\ -\ 2\cdot78 \\ \hline \\ \hline \end{array}$$

5. $$\begin{array}{r} \text{l} \\ 1\cdot375 \\ \times\ \ \ \ 4 \\ \hline \\ \hline \end{array}$$

6. $$\begin{array}{r} \text{kg} \\ 0\cdot787 \\ \times\ \ \ \ 6 \\ \hline \\ \hline \end{array}$$

7. kg $5\overline{)4\cdot360}$

8. m $8\overline{)9\cdot36}$

9. 2·36 m + 0·74 m + 3·74 m

10. 4·000 kg − 1·764 kg

11. 2·450 l × 7

12. 6·540 kg ÷ 2

13. 5·640 l ÷ 10

14. Find the difference between 3·540 kg and 2·225 kg.

15. What is the total of 3 l, 2·145 l and 0·364 l?

16. Divide 4·20 m by 5.

17. Multiply 3·426 kg by 2.

18. What is half of 3·420 l?

19. What is double 2·375 kg?

20. What is the total of A, B and D?

21. What is the difference between B and C?

22. How much heavier is D than B?

23. Which weight is twice as heavy as B?

A	3 kg
B	$2\frac{1}{4}$ kg
C	3·240 kg
D	4·5 kg

24. Simon weighs 56·5 kg and his brother Ian weighs 70 kg. How much heavier is Ian than Simon?

25. Kreel and Tapton are 24 km apart. Sarah leaves Kreel and travels three-quarters of the way to Tapton. How many km has she travelled?

26. Norman buys a 2 m length of timber. He cuts 143 cm from it. How much timber is left?

27. Mrs. Burton buys 5 l of vinegar. How many ml will be left when she has used three-quarters of the vinegar?

Volume

Interlocking plastic cm cubes 200 ml measuring cylinder

Pour water into the measuring cylinder to the 100 ml mark.
Fasten 20 cubes together.
Drop them into the cylinder.
All the cubes should be under the water.
By how many ml has the water level risen?

Take out the cubes.
Check the water level is still 100 ml.
Rearrange the cubes.
Drop them into the water.
By how many ml has the water level risen?

Repeat this with another arrangement of the cubes.
What do you notice?

Estimate how many ml the level will rise by if you drop in 30 cubes.
Check whether you were right.

Number

This table shows the number of spectators at a Horse Show.

Day	Attendance
Monday	812
Tuesday	792
Wednesday	1000
Thursday	881
Friday	1109
Saturday	1430

1. What was the total attendance at the show?
2. What was the average daily attendance?
3. Admission to the show was £2 per person. How much was collected in admission charges?

Records broken at Horse Show

The Aberdeen Horse Show was a great success again this year. Good weather helped to break attendance records during the week. The attendance each day was:

Monday	800
Tuesday	800
Wednesday	1000
Thursday	900
Friday	1100
Saturday	1400

The attendance figures shown in the report are approximate.
We do not always need to know exact numbers.
The reporter **rounded off** the figures to the nearest hundred.

4. What was the total attendance shown by the newspaper?

Rounding off

This diagram shows how to round off to the nearest hundred.

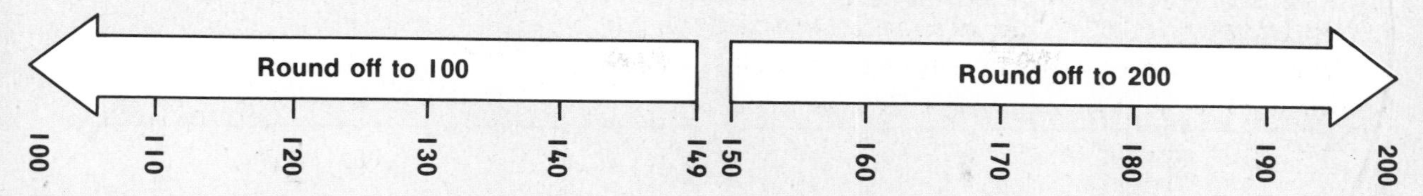

All the numbers from 100 to 149 are rounded off to 100.
All the numbers from 150 to 200 are rounded off to 200.

Round off these numbers to 100 or 200.

1. 121 **2.** 146 **3.** 193 **4.** 103 **5.** 176 **6.** 155
7. 180 **8.** 161 **9.** 183 **10.** 109 **11.** 150 **12.** 125

> When you round off, the "half-way number" is always rounded upwards.

Round off these numbers to the nearest hundred.

13. 107 **14.** 214 **15.** 286 **16.** 416 **17.** 394 **18.** 465
19. 511 **20.** 712 **21.** 391 **22.** 95 **23.** 873 **24.** 401
25. 1303 **26.** 1268 **27.** 1121 **28.** 1371 **29.** 2128 **30.** 1643

This table shows the number of spectators at a previous Horse Show.

Write a newspaper report about this Horse Show.
Remember to round off the attendance figures.

Day	Attendance
Monday	732
Tuesday	681
Wednesday	887
Thursday	610
Friday	950
Saturday	1208

Sometimes we round off to the nearest thousand.
It is the same rule as for hundreds.

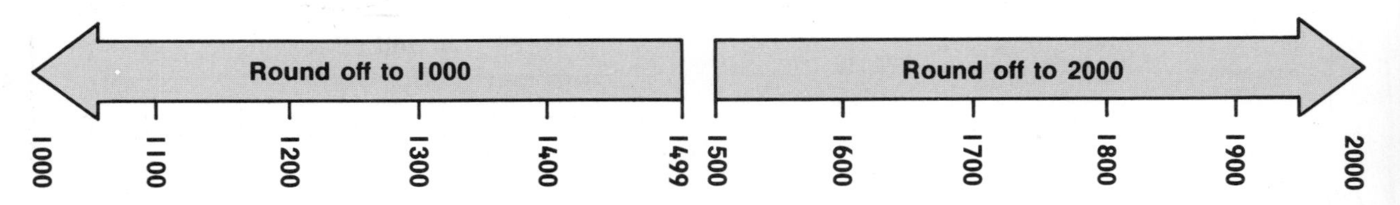

All the numbers from 1000 to 1499 are rounded off to 1000.
All the numbers from 1500 to 2000 are rounded off to 2000.

Round off these numbers to the nearest thousand.

1. 1218	2. 1462	3. 1938	4. 1003	5. 1764	6. 1551
7. 2493	8. 6858	9. 3576	10. 7496	11. 9037	12. 5111

By rounding off you can find approximate answers to check your working.
Complete these examples, and check your answers by rounding off.

13.

Example	**Rounded off**
1401	1000
2286	2000
+ 5721	+ 6000

14.

Example	**Rounded off**
1873	2000
− 949	− 1000

15.

Example	**Rounded off**
1963	2000
× 4	× 4

16.

Example	**Rounded off**
$4\overline{)3896}$	$4\overline{)4000}$

Choose the correct answer for each of these. Rounding off will help you.

17. 1787 + 2131 + 1206 = (1934, 2014, 5124)

18. 4995 − 1068 = (3927, 1927, 2927)

19. 1009 × 5 = (3045, 7045, 5045)

20. 5898 ÷ 3 = (1966, 1066, 2966)

This table shows the approximate number of people who live in some of the cities in Great Britain.

City	Population
Liverpool	748 000
Sheffield	494 000
Newcastle	270 000
Glasgow	960 000

City	Population
Belfast	412 000
Carlisle	72 000
Plymouth	247 000
Swansea	167 000

The numbers shown in the table are very large.
To work out the population of Liverpool, you need to put the number shown in the table into place value columns.

Hundred thousands	Ten thousands	Thousands	Hundreds	Tens	Units
7	4	8	0	0	0

The population of Liverpool is seven hundred and forty-eight thousand.

1. Write in words the populations of the other cities.
2. Which city has most people?
3. Which city has fewest people?
4. Write the names of the cities in order of size of population.

These numbers show the populations of other cities.
Write them in figures.

5. Two hundred and twenty-four thousand.
6. One hundred and eight thousand.
7. Three hundred and forty-two thousand.
8. Five hundred and thirty-three thousand.
9. One hundred and thirty-two thousand.
10. Ninety-seven thousand.

Plane shapes

These lines will never meet.
These lines are **parallel**.

To indicate that lines are parallel they are often marked with arrow heads.

Find these shapes.
Draw round them.
Mark the parallel sides.

Find a shape which has only one pair of parallel sides, like this:

Draw round it.

≫ It is a **trapezium**.
Mark its parallel sides.

Geo-strips, plane shapes

Make a square with geo-strips.

Write what you know about the length of the sides.
Which sides are parallel?
What do you know about the angles?

Twist it like this.
You have made a **rhombus**.
Are the sides still the same length?
Are the sides still parallel?
Are the angles still right-angles?
Find a rhombus in the shape box.
Draw round it and label it.
Write three properties of a rhombus.

Make a rectangle with geo-strips.
What do you know about the length of the sides?
What do you know about the angles?
Are any sides parallel?

Twist it like this.
You have made a **parallelogram**.

Find a parallelogram in the shape box.
Draw round it and label it.
Write three properties of a parallelogram.

Area

Calculate the shaded area.

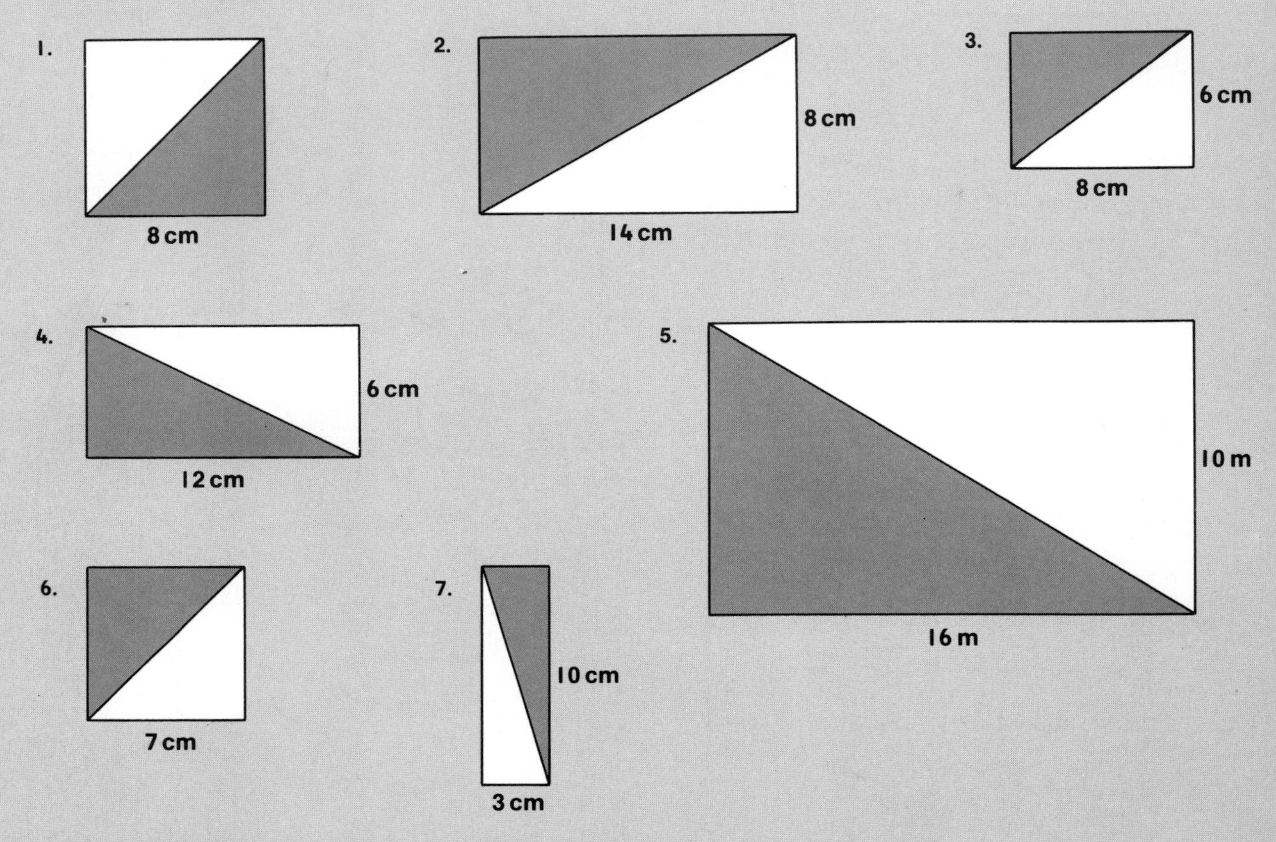

Calculate the areas of these right-angled triangles.

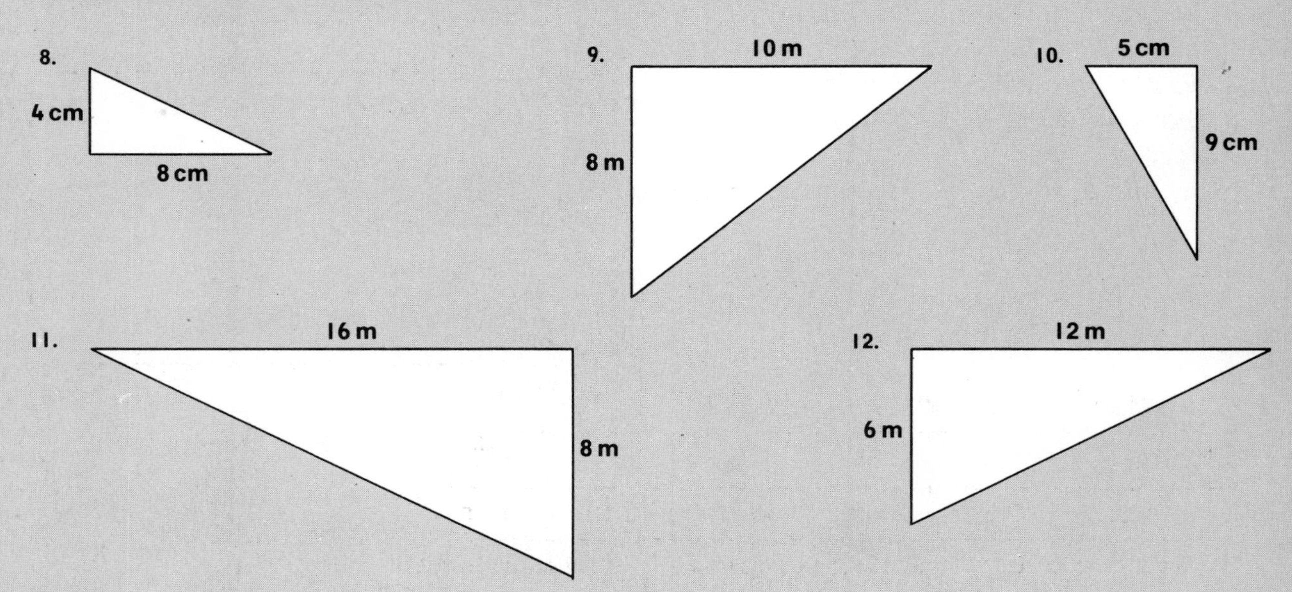

1. Find the area of this square.
Double the length of each side and find the area.
How many times greater has the area become?

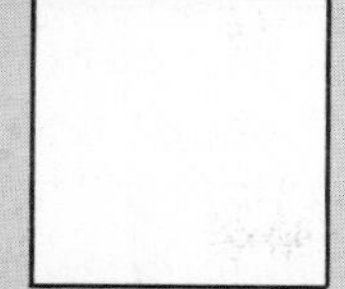

2. Find the area of this rectangle.
Double the length of each side and find the area.
How many times greater has the area become?

3. This table shows areas of rectangles.

Rectangle	A	B	C	D	E
Area	14 cm²	28 cm²	22 cm²	36 cm²	50 cm²

If the sides of each rectangle are doubled, what will its new area be?

4. Find the area of this square.
Halve the length of each side and find the area.
How many times smaller has the area become?

5. Find the area of this rectangle.
Halve the length of each side and find the area.
How many times smaller has the area become?

6. This table shows areas of rectangles.

Rectangle	A	B	C	D	E
Area	20 cm²	24 cm²	36 cm²	52 cm²	76 cm²

If the sides of each rectangle are halved, what will its new area be?

Percentages

> This sign **%** means **per cent**.
> "Per cent" means "out of a hundred".

Calair Travel

10% off all package tours to Greece

Write three words beginning with "cent".
Explain their meanings.

> 30 out of 100 can be written as 30%.

Complete these:

1. 50 out of 100 = ☐ %
2. 70 out of 100 = ☐ %
3. 90 out of 100 = ☐ %
4. 10 out of 100 = ☐ %
5. 80 out of 100 = ☐ %
6. 40 out of 100 = ☐ %

Write the meanings of these:

7. 70%	8. 40%	9. 100%	10. 25%	11. 80%	12. 95%
13. 10%	14. 75%	15. 50%	16. 60%	17. 5%	18. 35%

1. 32 out of 100 children said they did not like ice-cream. Write the percentage of children who disliked ice-cream.
2. In a road survey 64 out of every 100 vehicles were cars. Write the percentage of cars in the survey.
3. 16 out of 100 creatures, living in a zoo belonged to the cat family. Write the percentage of creatures belonging to the cat family.
4. 15 out of a group of 100 passengers had no hand luggage. Write the percentage of passengers with no hand luggage.

> 50 out of 100 can be written as $\frac{50}{100} = 50\%$.
>
> 75 out of 100 can be written as $\frac{75}{100} = 75\%$.

Write these as percentages.

5. $\frac{30}{100}$ 6. $\frac{90}{100}$ 7. $\frac{5}{100}$ 8. $\frac{60}{100}$ 9. $\frac{40}{100}$ 10. $\frac{10}{100}$

11. $\frac{70}{100}$ 12. $\frac{15}{100}$ 13. $\frac{85}{100}$ 14. $\frac{95}{100}$ 15. $\frac{20}{100}$ 16. $\frac{75}{100}$

> All numbers to be changed into percentages have to be out of 100.
>
> 25 out of 50 can be written as $\frac{25}{50} = \frac{50}{100} = 50\%$.
>
> 15 out of 25 can be written as $\frac{15}{25} = \frac{60}{100} = 60\%$.

Write these as percentages.

17. $\frac{40}{50}$ 18. $\frac{10}{25}$ 19. $\frac{20}{50}$ 20. $\frac{20}{25}$ 21. $\frac{7}{10}$ 22. $\frac{9}{10}$

23. $\frac{4}{5}$ 24. $\frac{2}{5}$ 25. $\frac{35}{50}$ 26. $\frac{15}{50}$ 27. $\frac{9}{20}$ 28. $\frac{18}{20}$

29. $\frac{3}{10}$ 30. $\frac{3}{5}$ 31. $\frac{5}{25}$ 32. $\frac{45}{50}$ 33. $\frac{30}{50}$ 34. $\frac{1}{5}$

Here are some spelling tests.

1. Write the numbers of correct spellings as percentages.

Nicola
rectangel ✗
entertain ✓
carraige ✗
tommorrow ✗
efficient ✓
assistant ✓
dungeon ✓
regular ✓
receive ✓
estimate ✓
7/10

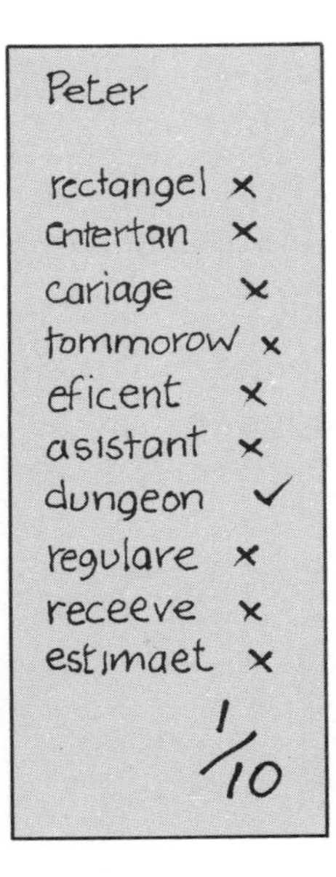

James
rectangle ✓
entertain ✓
carrege ✗
tomorow ✗
efficent ✗
assistent ✗
dungeon ✓
regular ✓
receve ✗
estimate ✓
5/10

Susan
rectangle ✓
entertain ✓
carriage ✓
tomorrow ✓
efficient ✓
assistant ✓
dungeon ✓
regular ✓
receive ✓
estimate ✓
10/10

Jane
rectangle ✓
entertain ✓
carriage ✓
tomorrow ✓
eficeint ✗
assistant ✓
dungeon ✓
regular ✓
receeve ✗
estimate ✓
8/10

2. 2 out of 5 children have blue eyes.
   Write the percentage of children who have blue eyes.
3. 7 children out of every 20 eat sandwiches at school.
   Write the number of children who eat sandwiches as a percentage.
4. 16 out of 25 children prefer ice-cream to lollies.
   What percentage of children prefer ice-cream?
5. 7 out of 10 crayons in a box are red.
   Write the percentage of red crayons in the box.

Write the fraction shaded as a percentage.

6.
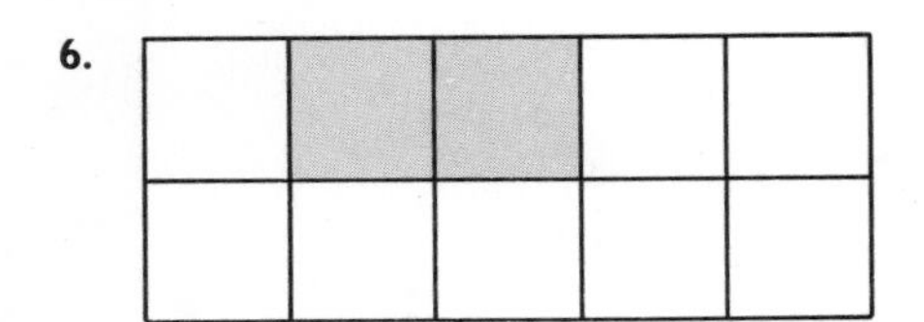

7.
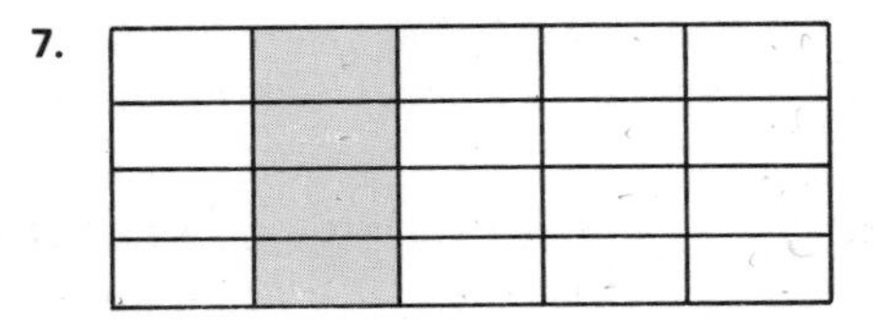

8.
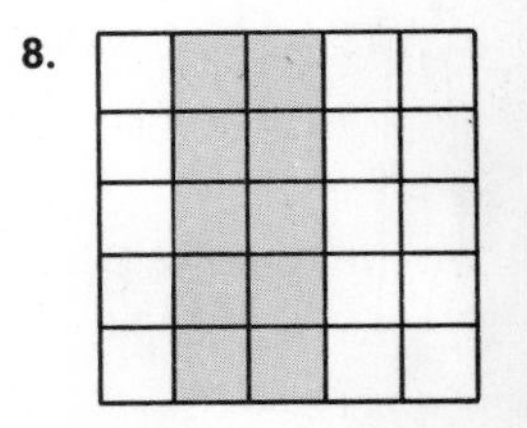

Write these as fractions in their simplest form.

9. 50%	10. 25%	11. 75%	12. 20%	13. 60%	14. 90%
15. 5%	16. 15%	17. 70%	18. 35%	19. 45%	20. 80%

Find 10% of £50
$= \frac{1}{10}$ of £50
= £5

Find 75% of £40
$= \frac{3}{4}$ of £40
= £30

Find:

1. 20% of £50
2. 40% of £25
3. 90% of £10
4. 80% of 50
5. 70% of 10
6. 80% of 25
7. 40% of £5
8. 90% of £20
9. 80% of £5
10. 90% of 30
11. 30% of 10
12. 75% of 200
13. 50% of £300
14. 80% of £500
15. 25% of £20

16. A man ordered a load of bricks costing £1000.
He paid a 30% deposit.
How much deposit did he pay?

17. A trawler captain unloaded his catch of 500 crates of fish.
Mr. Davids, the fish merchant, bought 60% of the catch.
How many crates did Mr. Davids buy?

18. A firm makes 2000 TV sets a week.
40% of them are exported.
How many are exported?

19. A shop sells 2500 tins of vegetables a month.
30% of the tins sold contain peas.
How many tins of peas are sold each month?

20. A garage sells 9000 litres of petrol each week.
Car owners buy 90% of the petrol.
How many litres do car owners buy each week?

21. Mr. James bought 160 kg of fertilizer for his garden.
His neighbour, Mr. Taylor, bought 20% as much.
How much fertilizer did Mr. Taylor buy?

22. A school library contained 3000 books.
70% of the books were fiction.
How many fiction books were in the library?

Graphs

Jackie and John are hikers.
They set off on a hike at 10 o'clock one morning.
The hike lasted 8 hours.
They had two rests, one for lunch and one for tea.
This graph shows how far they travelled.

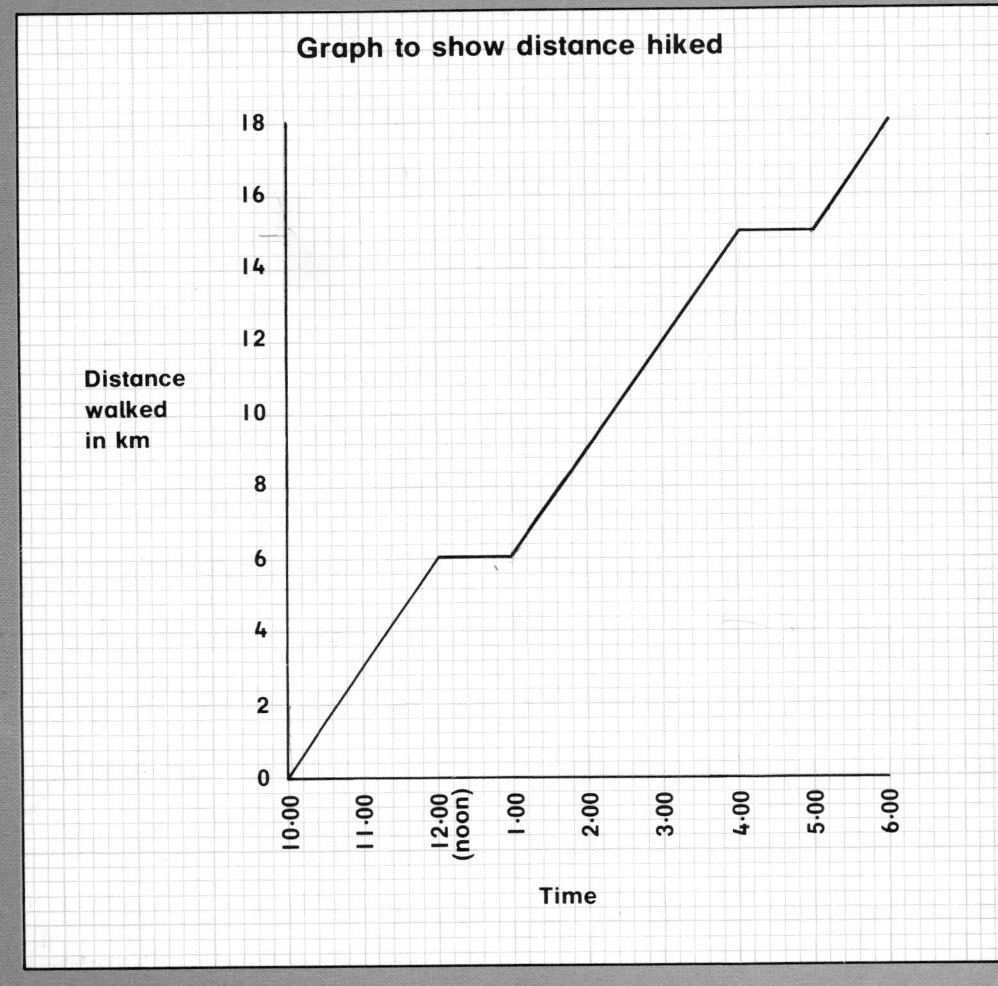

1. At what time did they stop for lunch?
2. How far did they walk before lunch?
3. At what time did they stop for tea?

4. How far did they walk between lunch and tea?
5. How far did they walk after tea?
6. How far did they walk altogether?
7. How much time did they spend resting?
8. How many hours did they spend walking?

Christine likes going on bike rides.
She travels at 8 km an hour when she cycles.
Yesterday she set out on a bike ride at 7.30 am.
This table shows how far she travelled each hour on her bike ride.

No. of hours	1	2	3	4	5	6	7	8	9
Distance in km	8	16	24	24	32	40	48	48	56

Draw a line graph to show the distance she travelled each hour.

From the graph find out:

9. the time she had the first rest.
10. the time she had the second rest.
11. at what time she had travelled 44 km.
12. at what time she had travelled 52 km.
13. how many km she had travelled by 11.30 am.
14. how many km she had travelled by 2.30 pm.
15. at what time she had travelled half way.

Volume

Remember: volume of a cuboid = area of base × height

Calculate the volumes of these cuboids.

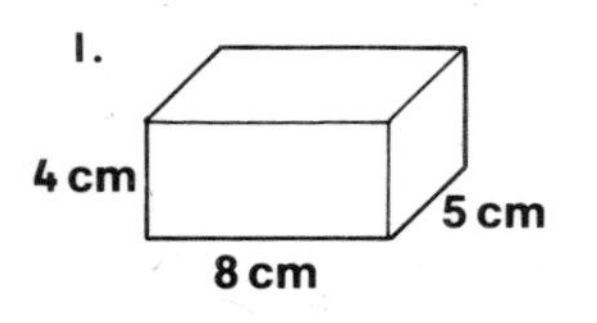

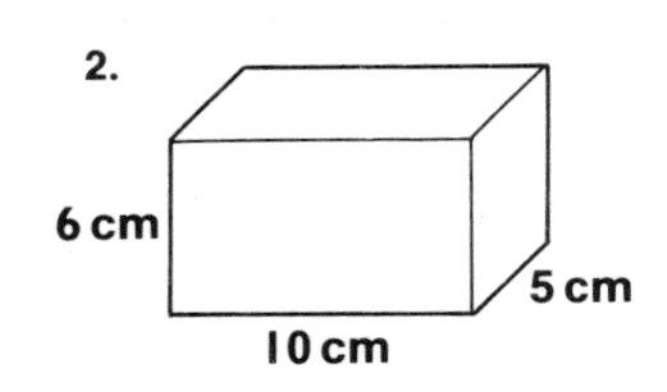

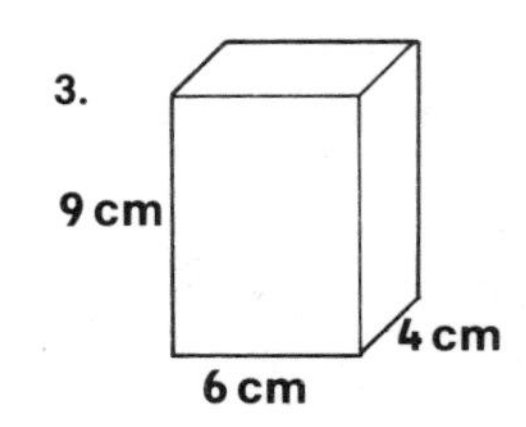

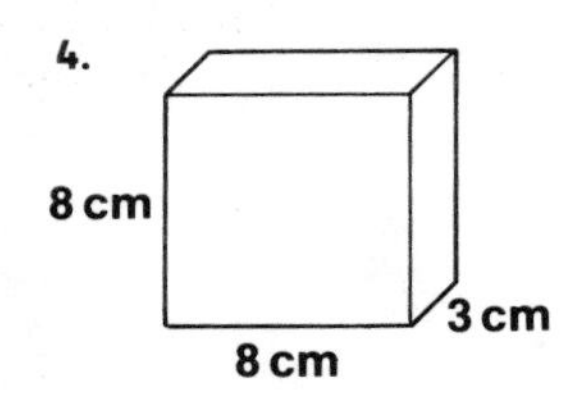

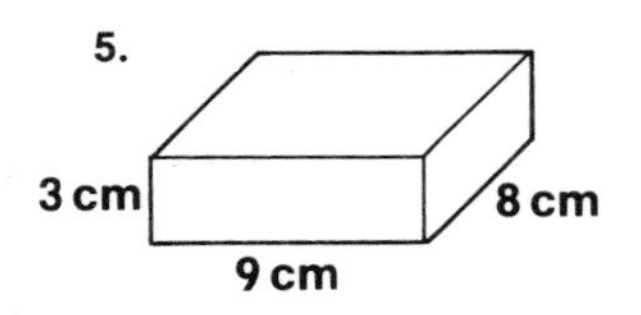

6. The table shows the sizes of cuboids.
Copy and complete it.

Length	Width	Height	Volume
4 cm	5 cm	6 cm	* cm^3
* cm	3 cm	4 cm	24 cm^3
4 cm	* cm	10 cm	80 cm^3
3 cm	3 cm	* cm	117 cm^3

7. Which of these cuboids have the same volume?

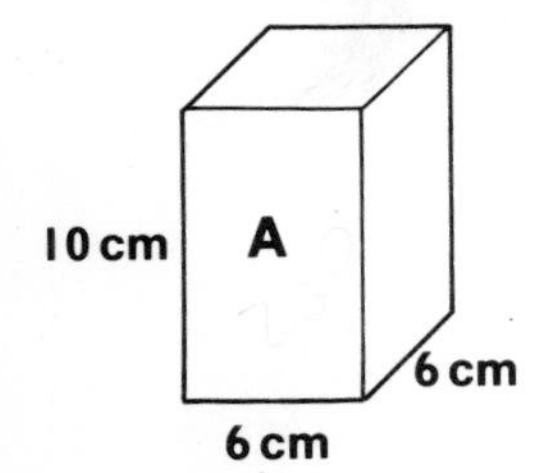

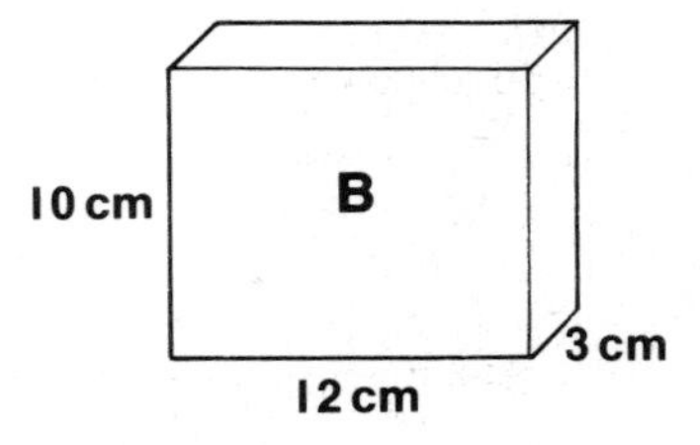

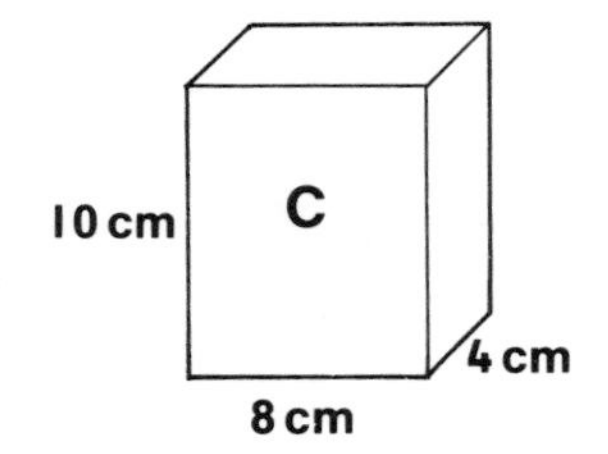

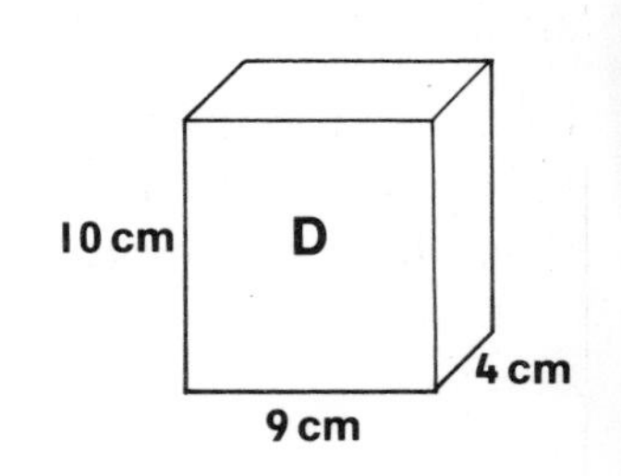

8. How many of these 2 cm cubes will fit into these boxes?

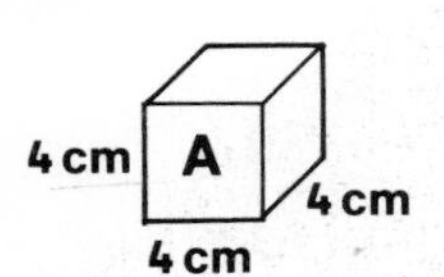

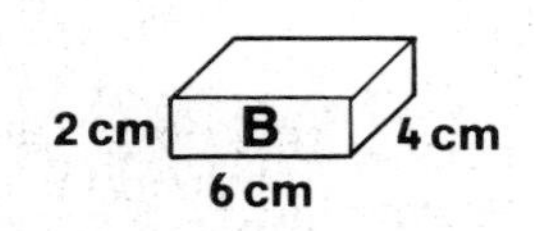

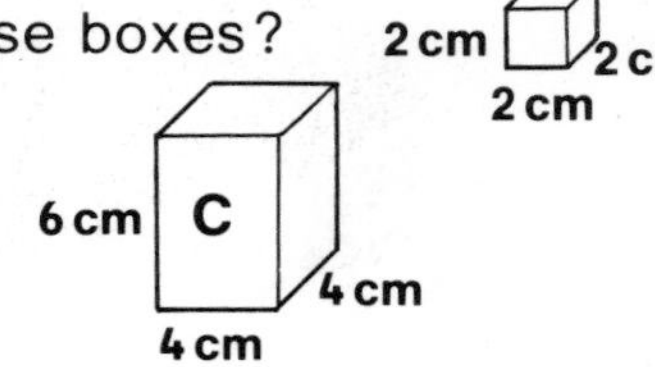

Measurement

100 sheets of paper, ball of string, strip of card, beans, scales

Measuring small things

How thick is paper?

Count 100 sheets.
Measure the thickness of the 100 sheets.
Calculate the thickness of 1 sheet.

How thick is string?

Wind string 10 times round a strip of card.
Measure the 10 thicknesses of string.
Calculate the thickness of the string.

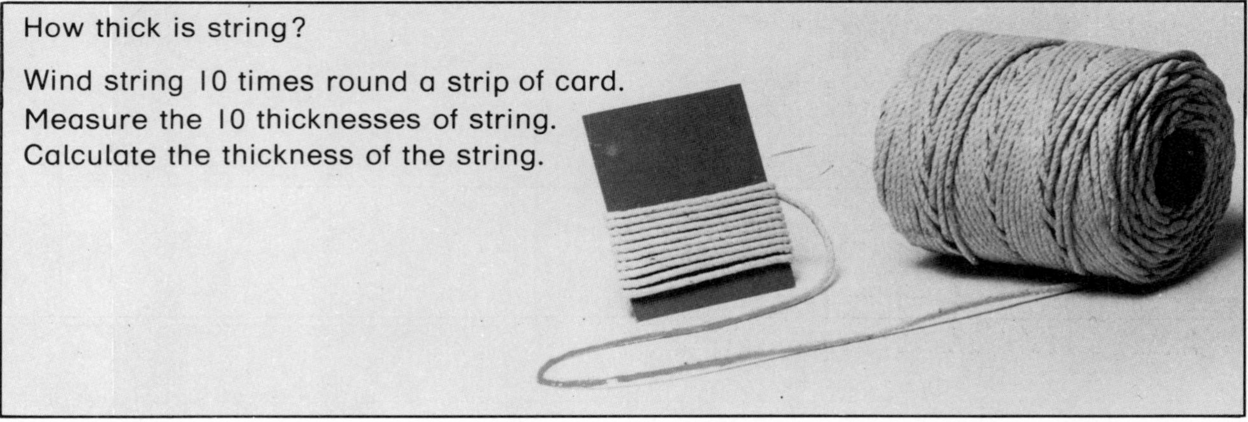

How heavy is a bean?

Count 100 beans.
Weigh the 100 beans.
Calculate the weight of 1 bean.

Area

1. Both sides of each panel must be painted.
   Each tin of paint covers 18 m².
   How many panels will one tin of paint cover?

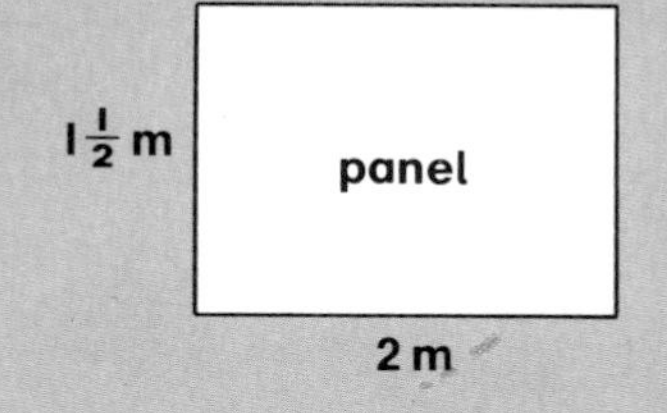

2. The paint costs £4·75 a tin.
   What will it cost to paint 12 panels?

3. Each tile measures 20 cm × 20 cm.
   How many tiles will cover 1 m²?

4. How many tiles will be needed to cover the wall?

5. Five tiles cost £4. What will it cost to tile the wall?

2 m

wall

1 m

2 m

1 m

6. Each roll is 50 cm wide and 6 m long.
   How many rolls of paper will be needed to paper the wall?

7. The paper costs £3·20 a roll. What change will there be from £10?

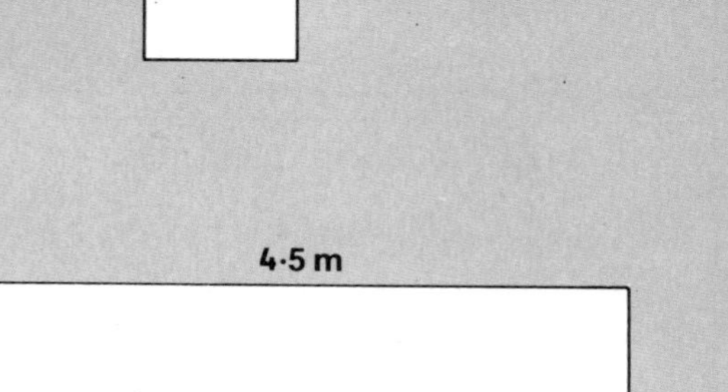

Shape

Plane shapes

Find a parallelogram.
Draw round it.
Rotate the parallelogram.

Does it fit into the outline again before it completes a full turn?

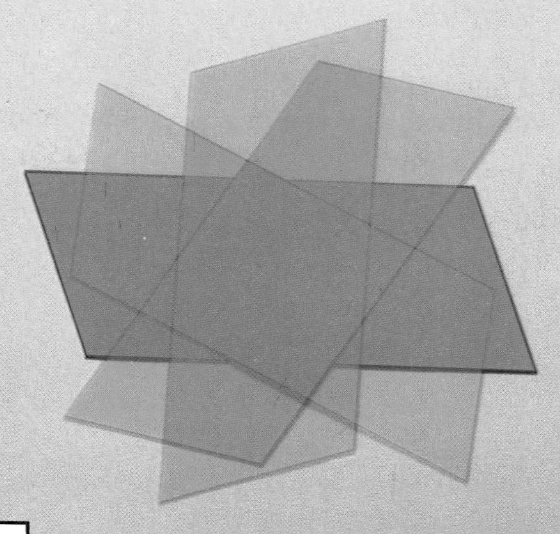

A shape that fits into its own outline during a rotation has **rotational symmetry**.

Can you find any other shapes that have rotational symmetry?
Draw round them.

≫ These shapes have rotational symmetry.

Write which shapes have rotational symmetry, line symmetry, or both.

1.

2.
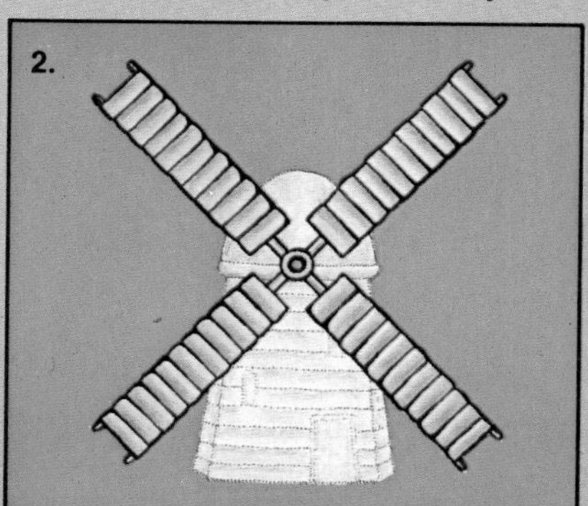

3.

4.
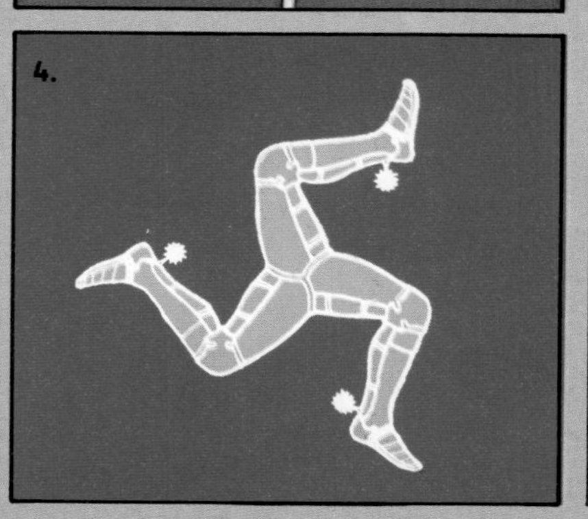

5.
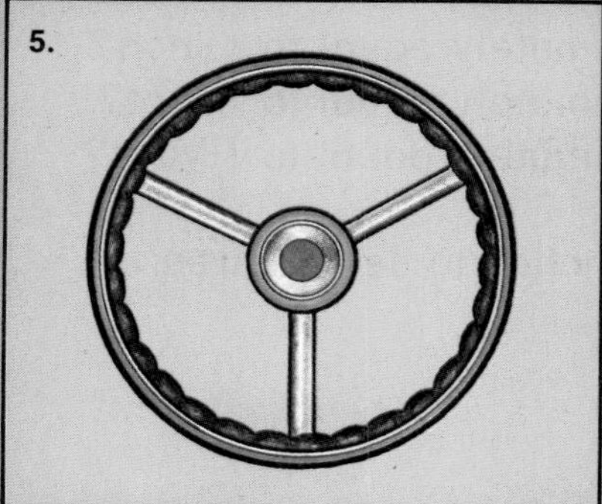

6.
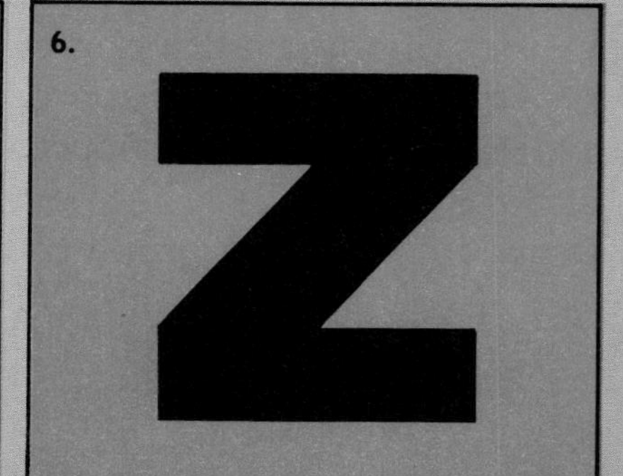

Measurement

Imperial units

When your parents were at school they had to measure using **inches**.

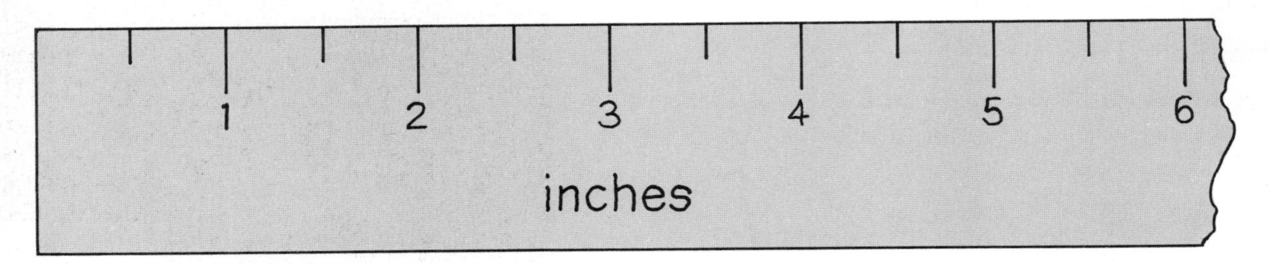

Make a paper strip which is 12 inches long.
Use the guide above to help you.
Mark in the inches and half inches.
Using your paper strip, measure and record the lengths of six items to the nearest half inch.

A length of 12 inches is called a **foot**.

⪢ 12 inches = 1 foot
12 ins = 1 ft

Using your paper strip, measure and record the lengths of six items to the nearest foot.

A length of 3 feet is called a **yard**.

⪢ 3 feet = 1 yard
3 ft = 1 yd

Using your paper strip to help you, cut a piece of string 1 yard long.
With your yard of string, measure and record the lengths of six items to the nearest yard.

1. How many cm are approximately equal to 1 inch?
2. How many cm are approximately equal to 1 foot?
3. How many cm are approximately equal to 1 yard?

Draw a conversion graph of inches to centimetres.

Investigations

Calculator

All at sixes and sevens

Use a calculator to work these out.

$143 \times 7 \times 1$

$143 \times 7 \times 2$

$143 \times 7 \times 3$

$143 \times 7 \times 4$

What do you notice?
Does the pattern continue?

$15873 \times 7 \times 1$

$15873 \times 7 \times 2$

$15873 \times 7 \times 3$

$15873 \times 7 \times 4$

What do you notice?
Does the pattern continue?

6×7

66×67

666×667

6666×6667

What do you notice?
Does the pattern continue?

142857×1

142857×2

142857×3

142857×4

What do you notice?
Does the pattern continue?

Two coins

Heads or tails?
If you spin a coin, which is more likely to turn up, heads or tails?

Spin a coin 50 times.
Keep a tally of your results.

Which turned up more times, heads or tails?
Will this always happen?

		Total
Heads	~~IIII~~ III	
Tails	~~IIII~~ ~~IIII~~	

If you spin two coins, which combination is most likely to turn up:

a) two heads?
b) two tails?
c) one head and one tail?

Spin two coins 50 times.
Keep a tally of your results.

Which combination turned up most?
Why was this so?

		Total
Two heads		
Two tails		
Head and tail		

Two dice

When you roll a die, which number is most likely to turn up?

Roll a die 50 times.
Keep a tally of numbers turning up.

Which number turned up most?
Will this always happen?

Number on die		Total
1		
2		
3		
4		
5		
6		

When you roll two dice, which total do you think occurs most often?

Roll two dice 50 times.
Keep a tally of the totals.

Which total turned up most?
Why was this so?

Total on dice		Total
2		
3		
4		
5		
6		
7		
8		
9		
10		
11		
12		

These two straight lines cross at one point.

Can you draw two straight lines to cross at more than one point?

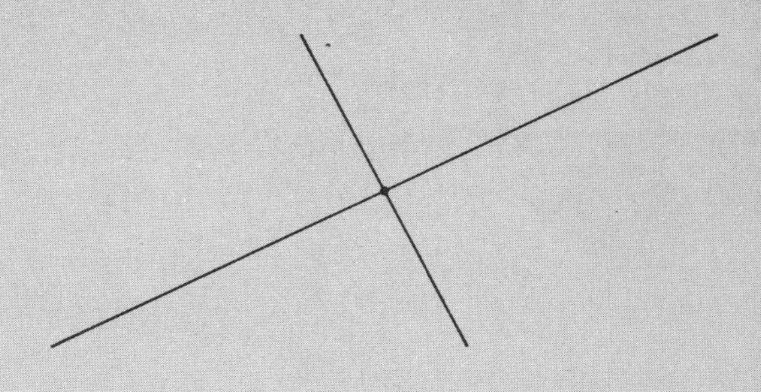

These three straight lines cross at two points.

Can you draw three straight lines to cross at more than two points?

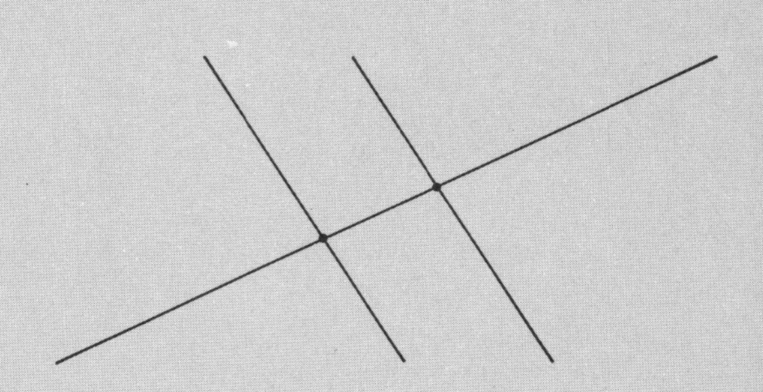

Draw four straight lines to give the maximum number of crossing points.

Can you calculate the maximum number of crossing points you can get with:

a) five lines?
b) eight lines?
c) ten lines?

A region is an enclosed area.
Three straight lines can cross to form a region.

What is the maximum number of regions you can make with:

a) four lines?
b) seven lines?
c) nine lines?

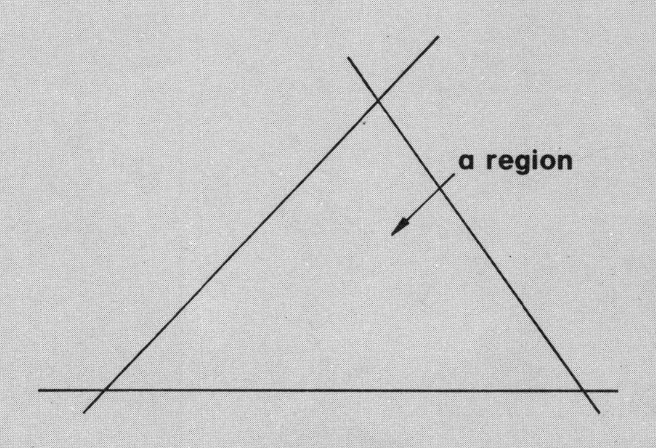

Squared paper

Draw this shape on squared paper.
Cut it out.
Cut it into two parts and rearrange the parts to make a square.
Stick the square into your book.

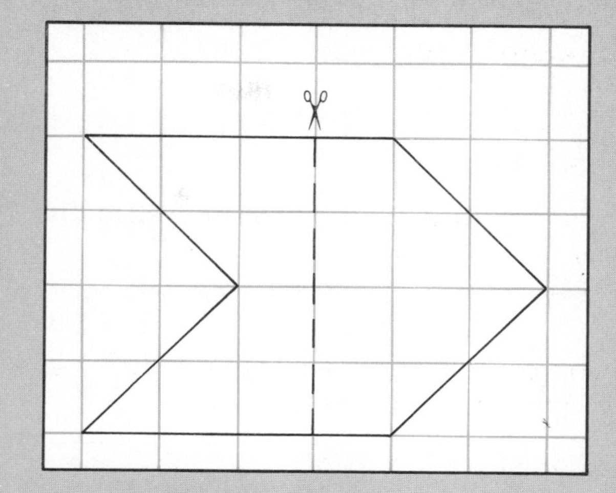

Do the same with these shapes.

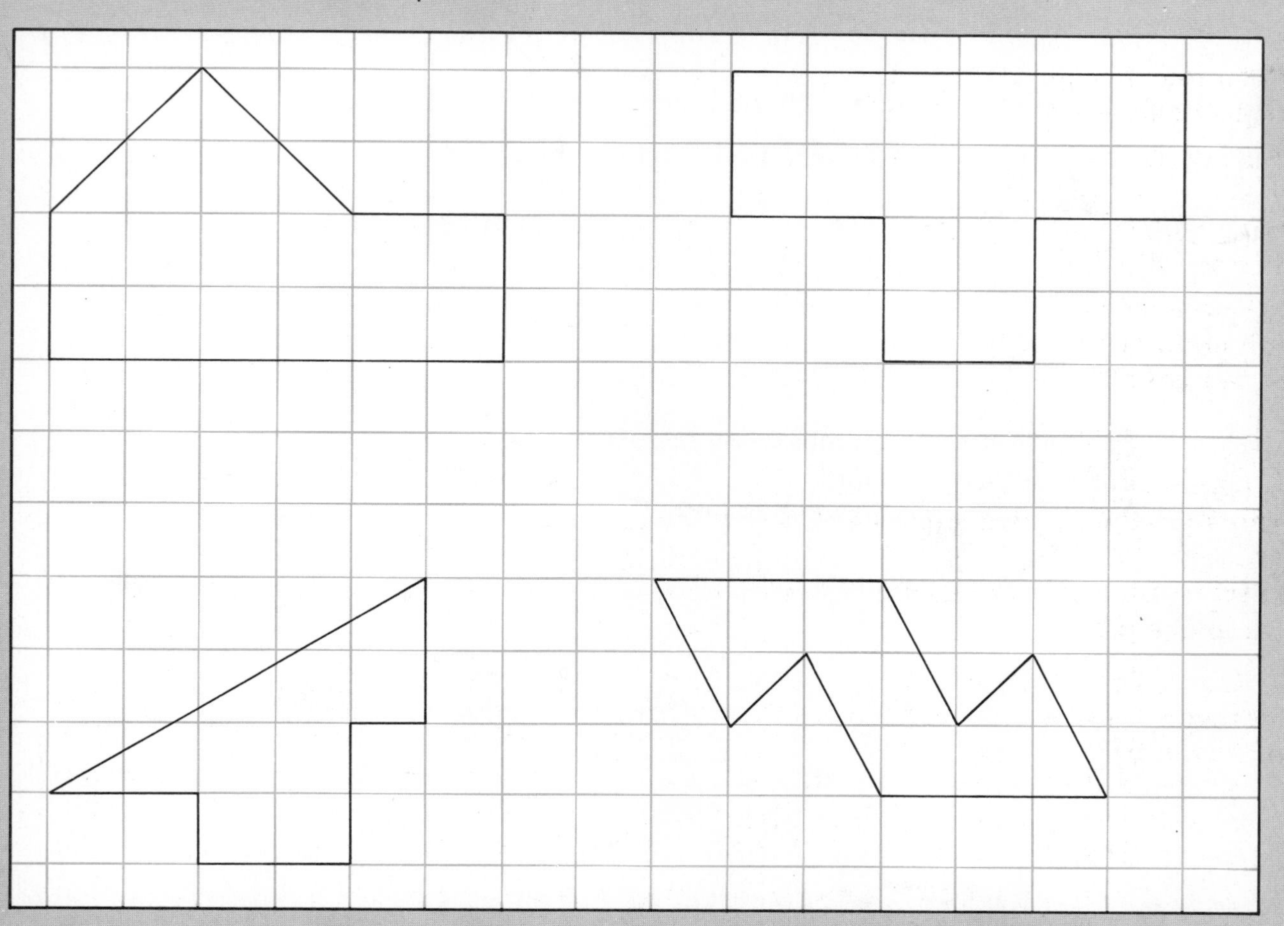

Assessment

1. Add 247, 1038 and 2792.
2. From 4026 subtract 1874.
3. Multiply £7·64 by 8.
4. Divide £21·78 by 9.
5. Find the total of 2·64 m, 3·92 m and 74 cm.

6. kg
$$\begin{array}{r} 1\cdot280 \\ +\ 3\cdot745 \\ \hline \\ \hline \end{array}$$

7. l
$$\begin{array}{r} 3\cdot750 \\ -\ 1\cdot855 \\ \hline \\ \hline \end{array}$$

8. kg
$$\begin{array}{r} 3\cdot465 \\ \times\ \ \ \ 7 \\ \hline \\ \hline \end{array}$$

9. l
$$6\overline{)7\cdot560}$$

Write the answers in the simplest form.

10. $2\frac{2}{3} + 1\frac{1}{4}$

11. $4\frac{1}{2} - 1\frac{5}{6}$

12. Round these off to the nearest thousand. 1728, 2136, 4962, 3500

13. Write in figures: Three hundred and sixteen thousand.

14.
$$\begin{array}{r} 164 \\ \times\ \ 17 \\ \hline \\ \hline \end{array}$$

15. £
$$\begin{array}{r} 1\cdot28 \\ \times\ \ \ 21 \\ \hline \\ \hline \end{array}$$

16. Find 20% of £18·35.

17. John has 1404 foreign stamps.
25% are American stamps.
How many American stamps has he?

18. Calculate the missing angle.

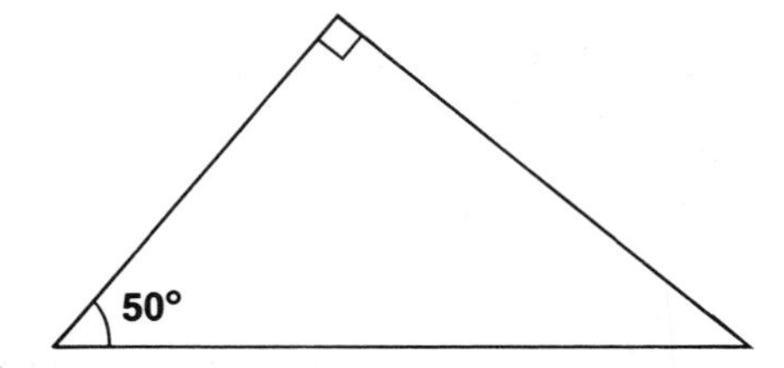

19.
$$\begin{array}{r} 7\cdot284 \\ +\ 1\cdot596 \\ \hline \\ \hline \end{array}$$

20.
$$\begin{array}{r} 6\cdot250 \\ -\ 2\cdot865 \\ \hline \\ \hline \end{array}$$

21. Write the value of the underlined digit. $2{\cdot}7\underline{4}2$

22. Find the area of this right-angled triangle.

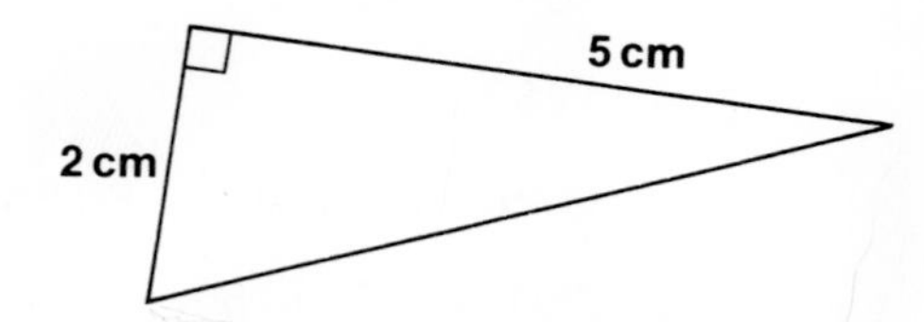

23. What is the area of this plot of land? Scale 1 cm : 5 m

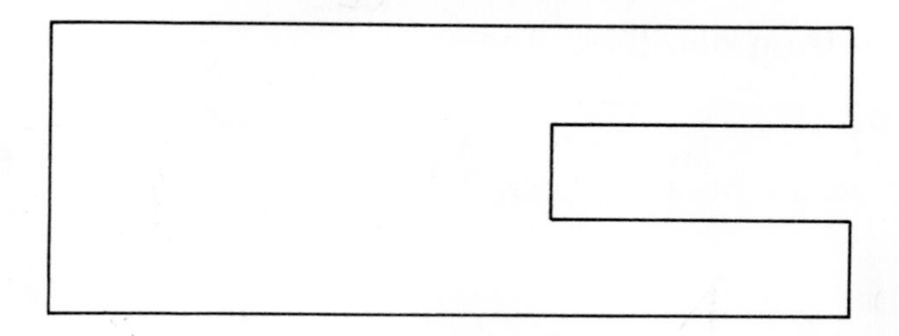

24. How much longer than $7\frac{1}{2}$ m is 9·45 m?

25. Name the shapes with parallel sides.

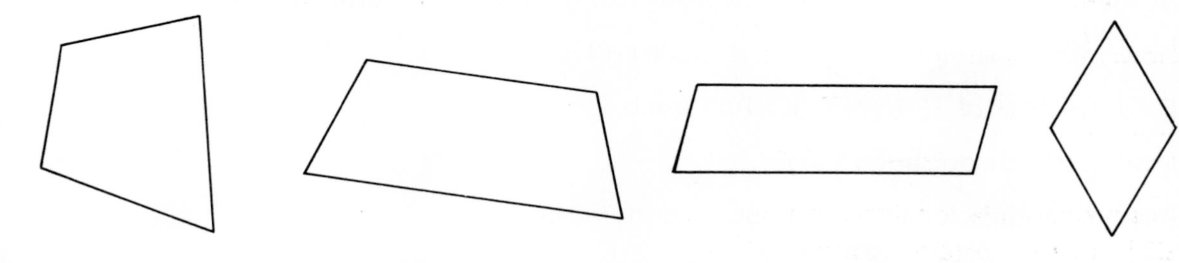

26. How many minutes are there from 0815 to 1035?

27. Calculate the volume of these cuboids.

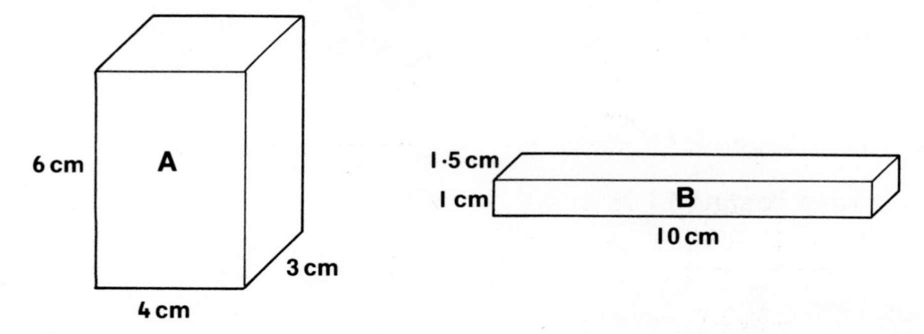

Number

Look for an easy way to do these.

Write the answers only.

1. $10 + 74 + 90$	2. $120 + 44 + 80$	3. $765 + 99$
4. $480 - 199$	5. 184×10	6. $1950 + 100$
7. $85 + 99 + 1$	8. $63 + 100 - 99$	9. £0·64 + £1·40 + £0·36
10. $2006 - 101$	11. $127 + 49 + 51$	12. $250 + 73 + 50$

13. To four thousand and seventy-eight add nine hundred and ninety-four.
14. Subtract eight hundred and sixty-nine from one thousand.
15. How much less than $1\frac{1}{2}$ kg is 0·650 kg?
16. Which amount of money is three times as much as £7·85?
17. Find the product of 27 and 148.
18. What weight is $\frac{2}{3}$ of 7·230 kg?
19. How much more than $2\frac{1}{2}$ l is 4·300 l?
20. Find $\frac{1}{2}$ of $7\frac{1}{2}$ l.
21. How much change will there be from £20 if I spend £14·55?
22. How much heavier than $4\frac{1}{2}$ kg is 5·300 kg?
23. Find the cost of 12 books at £1·55 each.
24. Make 2628 three times larger.
25. Find the average of $3\frac{1}{2}$ kg, 1·750 kg and 750 kg.
26. Multiply 36 by 25.
27. What is $\frac{3}{4}$ of £7·40?
28. Find the difference between 4·990 kg and 5·780 kg.
29. What must 9 be multiplied by to give 432?
30. Multiply £1·24 by 18.
31. Make 1·41 m three times smaller.
32. Find the cost of 12 books costing £1·36 each.
33. Find $\frac{4}{5}$ of 3 kg.
34. What is 3082 minus 1993?

All these cars were bought six months ago.

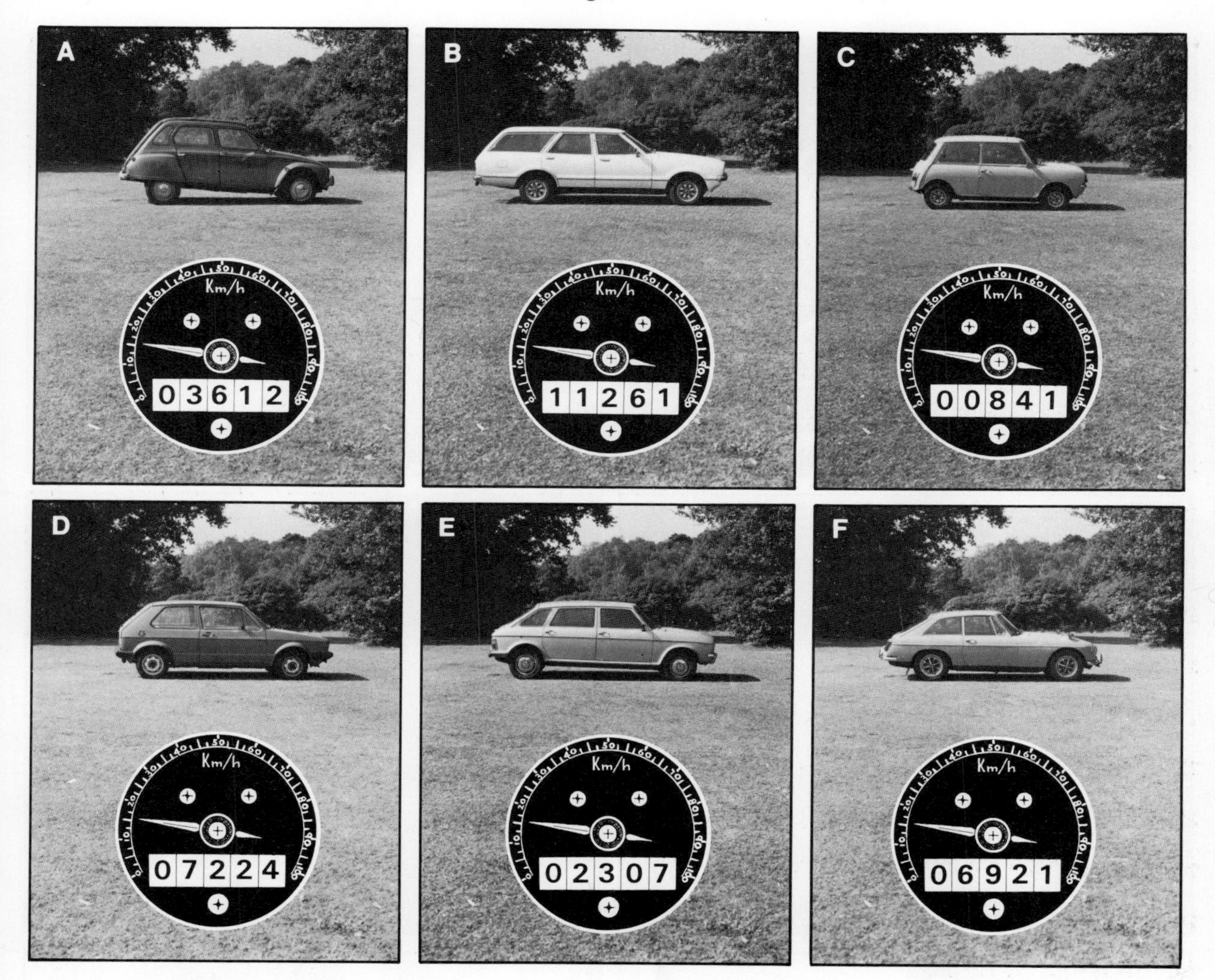

The dials show how many kilometres each car has travelled in the six months.

1. Which car has travelled furthest?
2. Which car has done the least number of km?
3. How much further has car A travelled than car E?
4. What is the difference in km travelled between cars B and F?
5. Which car has travelled twice as many km as car A?
6. Which car has travelled $\frac{1}{3}$ of the km of car F?
7. What is the total number of km travelled by all the cars?

Measurement

Write the measurement shown on each of the following:

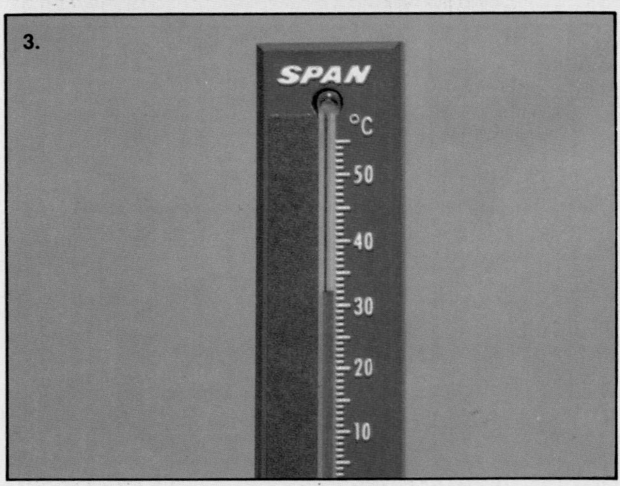

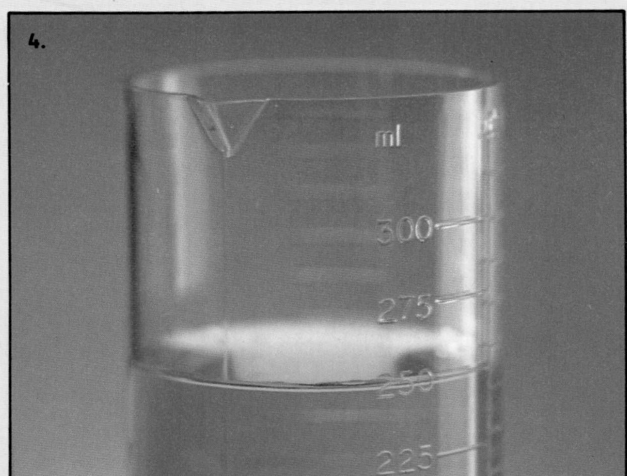

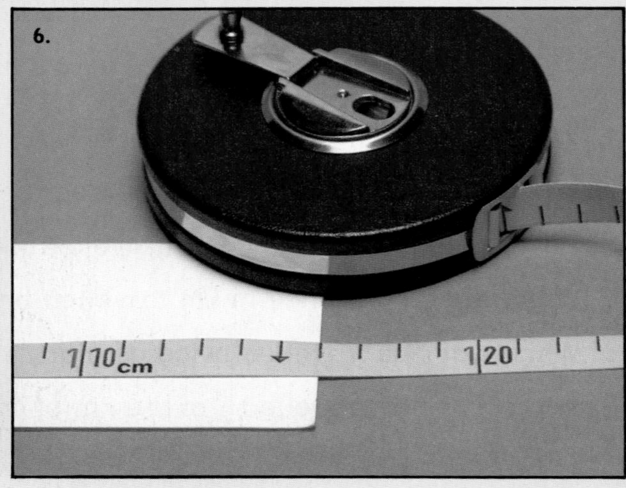

WAYMASTER
7.

GREAT BRITAIN
8.

C
SATCHWELL
9.

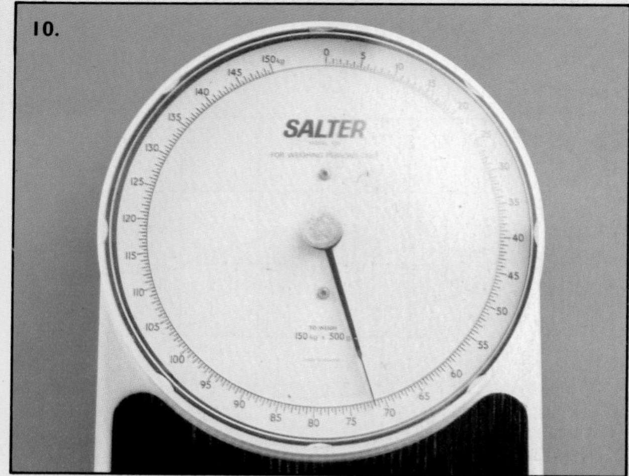
SALTER
10.

11.

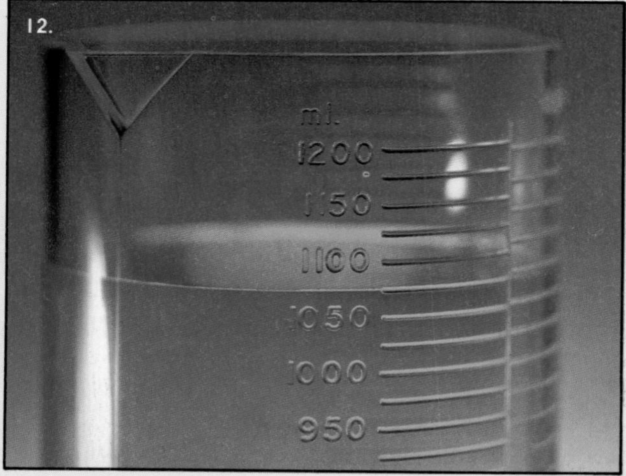
ml
1200
1150
1100
1050
1000
950
12.

Fractions

Write the fraction shaded, in its simplest form.

1.

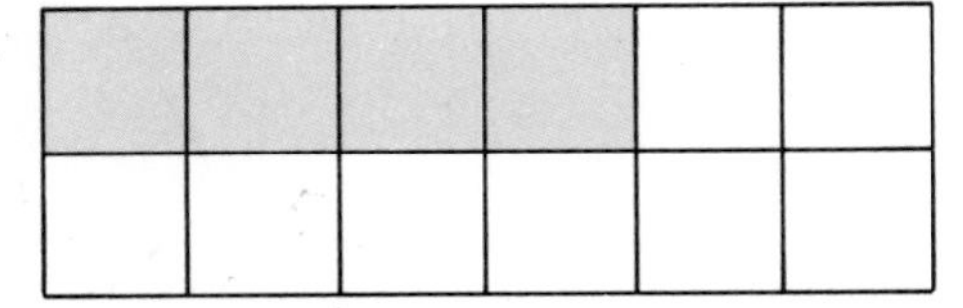

2.

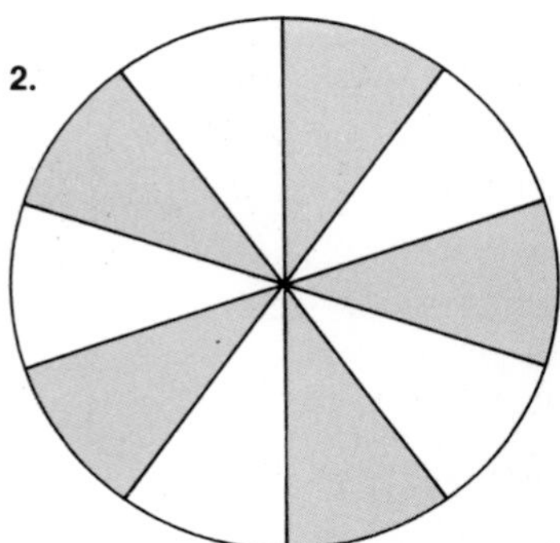

3.

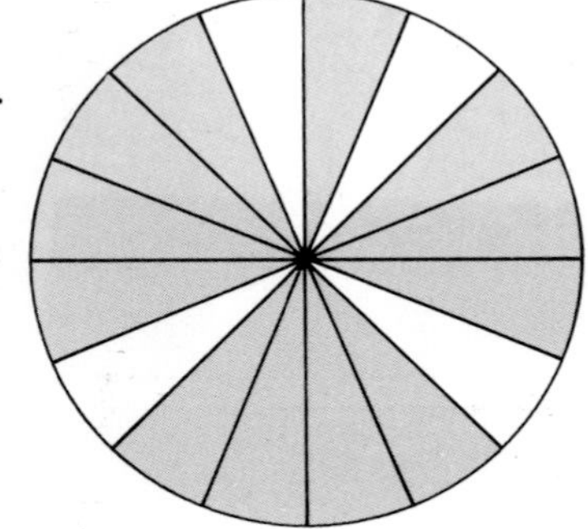

4.

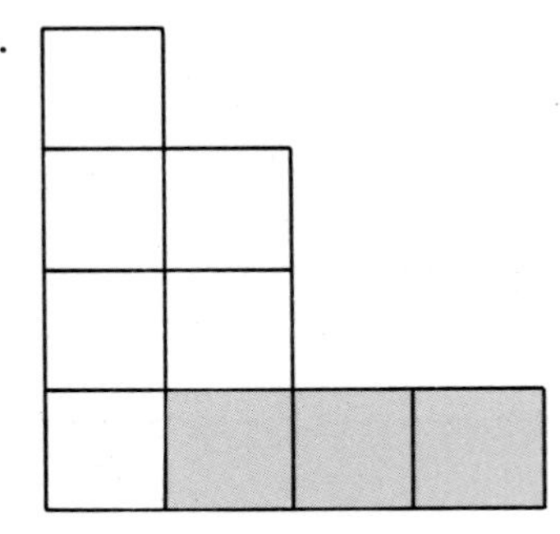

5.

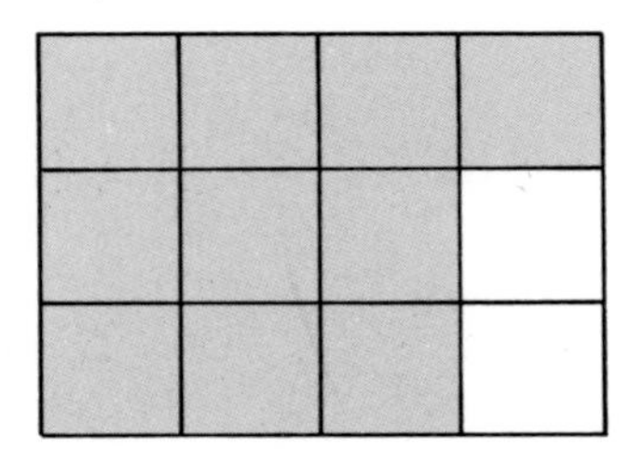

6.

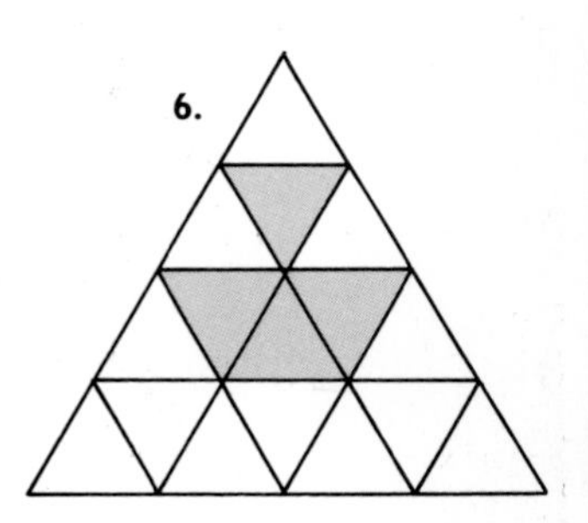

Complete these:

7. $\frac{1}{3} = \frac{*}{9}$ 8. $\frac{1}{4} = \frac{*}{16}$ 9. $\frac{*}{3} = \frac{3}{9}$ 10. $\frac{2}{3} = \frac{*}{12}$ 11. $\frac{*}{4} = \frac{12}{16}$

Write these fractions in order, smallest first.

12. $\frac{3}{5}, \frac{5}{6}, \frac{1}{2}$ 13. $\frac{5}{12}, \frac{1}{2}, \frac{1}{4}$ 14. $\frac{5}{6}, \frac{7}{8}, \frac{7}{12}$ 15. $\frac{11}{16}, \frac{3}{4}, \frac{1}{2}$

16. $\frac{1}{4}, \frac{1}{6}, \frac{5}{12}$ 17. $\frac{5}{6}, \frac{7}{9}, \frac{2}{3}$ 18. $\frac{1}{3}, \frac{1}{2}, \frac{5}{6}$ 19. $\frac{2}{5}, \frac{7}{10}, \frac{3}{4}$

Reduce these fractions to their simplest form.

20. $\frac{12}{16}$ 21. $\frac{18}{20}$ 22. $\frac{9}{15}$ 23. $\frac{6}{10}$ 24. $\frac{3}{9}$ 25. $\frac{14}{16}$

26. $\frac{20}{24}$ 27. $\frac{15}{18}$ 28. $\frac{5}{20}$ 29. $\frac{7}{21}$ 30. $\frac{4}{14}$ 31. $\frac{12}{15}$

Anne looked at some of her spelling tests.
On one test she had 20 correct out of 25.
The fraction she had correct was $\frac{20}{25}$.
This fraction written in its simplest form is $\frac{4}{5}$.
Anne had $\frac{4}{5}$ of the spellings correct.

Write these as fractions in their simplest form.

1. 16 out of 20 **2.** 18 out of 24 **3.** 20 out of 36

4. 25 out of 50 **5.** 15 out of 40 **6.** 25 out of 30

Do these. Write each answer in its simplest form.

7. $1\frac{1}{2} + 3\frac{2}{3}$ **8.** $5\frac{3}{8} + 2\frac{1}{3}$ **9.** $3\frac{3}{4} + 1\frac{4}{5}$ **10.** $\frac{7}{8} + 2\frac{1}{2} + 1\frac{5}{6}$

11. $1\frac{5}{12} + 2\frac{5}{6} + 1\frac{1}{4}$ **12.** $4\frac{3}{7} + 1\frac{1}{3}$ **13.** $\frac{1}{2} + 3\frac{4}{9} + 2\frac{1}{3}$ **14.** $2\frac{4}{5} + \frac{5}{6}$

15. $1\frac{7}{10} + 3\frac{3}{4} + 2\frac{1}{2}$ **16.** $2\frac{2}{3} + 3\frac{3}{10} + \frac{5}{6}$ **17.** $\frac{7}{12} + 2\frac{2}{3} + 3\frac{1}{2}$ **18.** $3\frac{1}{6} + 2\frac{7}{9} + \frac{1}{3}$

19. $3\frac{1}{2} - 1\frac{4}{9}$ **20.** $1\frac{1}{6} - \frac{4}{5}$ **21.** $4\frac{3}{4} - 3\frac{9}{10}$ **22.** $4\frac{3}{10} - \frac{1}{3}$

23. $5\frac{1}{6} - 3\frac{7}{9}$ **24.** $4\frac{1}{5} - 1\frac{3}{4}$ **25.** $6\frac{3}{8} - 4\frac{5}{6}$ **26.** $2\frac{1}{6} - 1\frac{1}{4}$

27. $3\frac{1}{3} - 1\frac{4}{7}$ **28.** $4\frac{5}{6} - 1\frac{5}{8}$ **29.** $5\frac{3}{4} - 2\frac{3}{8}$ **30.** $4\frac{2}{5} - 1\frac{5}{6}$

Write the fraction shaded in its simplest form.

31.

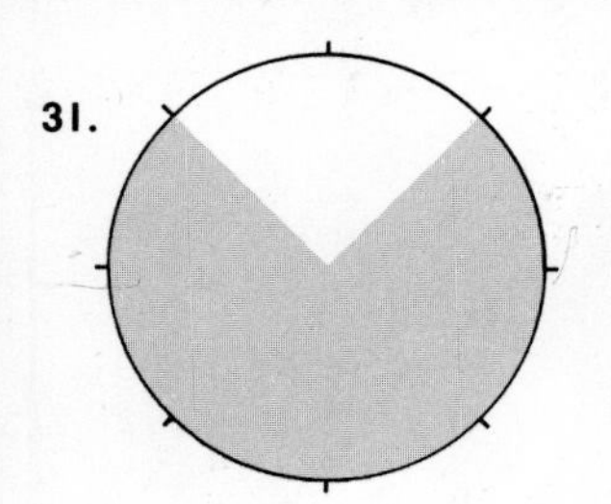

32.

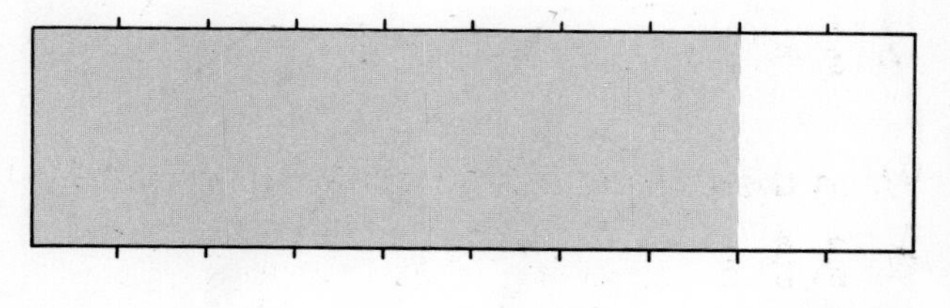

33.

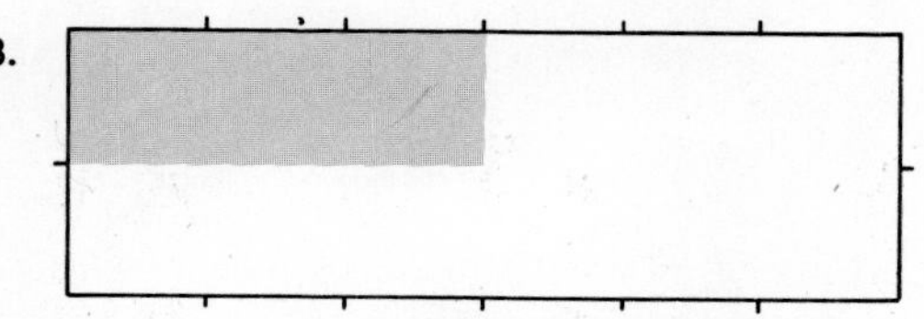

34.

Number

1. 329 + 18 + 4268
2. 5317 − 2986
3. £6·29 + £0·18 + £1·35
4. £8·63 − £5·92
5. 626 + 1298 − 373
6. £5·84 + £2·59 − £1·87
7. 4·35 m + 4·79 m
8. 7 m − 5·45 m
9. 3·25 m + 1·78 m − 0·95 m

10. Add 5 kg, 2·750 kg and $3\frac{3}{4}$ kg.
11. Add $1\frac{1}{2}$ l, 4·250 l and 0·790 l.
12. How much heavier than $2\frac{1}{2}$ kg is 4·250 kg?
13. What is the difference in weight between 4·520 kg and 3·755 kg?
14. How much further is $6\frac{1}{2}$ m than 1·75 m?
15. By how much is 4·750 l less than $9\frac{1}{2}$ l?
16. Find the total of 2·725 l, 550 ml and 6·750 l.

17. Find the total cost.
    Find the total weight.

18. Find the differences in cost.
    Find the differences in capacity.

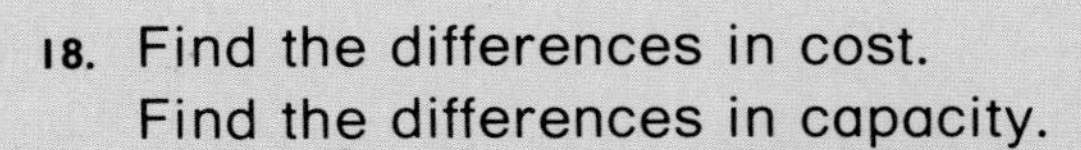

1. 128 × 6	2. 1136 × 7	3. 184 × 10	4. 1629 × 4
5. 164 × 12	6. 189 × 16	7. 212 × 23	8. 105 × 28
9. £1·78 × 5	10. £0·79 × 9	11. £3·25 × 4	12. £7·93 × 8
13. £1·36 × 13	14. £2·48 × 15	15. £1·72 × 21	16. £3·15 × 19

17. kg 3·740 × 6

18. l 2·955 × 7

19. kg 7·885 × 9

20. m 4·65 × 8

21. Multiply 7·585 l by 5.

22. Find the product of 236 and 18.

23. What weight is 5 times heavier than 2·750 kg?

24. What distance is 4 times further than 7·490 km?

25. 6314 ÷ 7	26. 4032 ÷ 8	27. 6939 ÷ 9	28. 4314 ÷ 6
29. £3·75 ÷ 5	30. £18·36 ÷ 4	31. £25·56 ÷ 9	32. £24·71 ÷ 7

33. kg 3) 4·650

34. kg 8) 7·224

35. l 6) 1·566

36. l 9) 7·848

37. kg 10) 6·950

38. l 7) 5·985

39. m 2) 6·58

40. m 8) 7·44

41. Divide £27·55 into 5 equal amounts.

42. Find the average of 7·36 m, 2·98 m and 91 cm.

43. Find $\frac{2}{5}$ of $7\frac{3}{4}$ kg.

44. Share 5·640 l into 4 equal quantities.

45. What amount is $\frac{3}{8}$ of £35·60?

46. Find $\frac{4}{5}$ of £17.

47. Divide 3·200 kg by 8.

Decimals

Multiply each of these numbers by 10.

1. 1·8	**2.** 0·67	**3.** 17·4	**4.** 0·08	**5.** 174
6. 0·009	**7.** 12·05	**8.** 24·6	**9.** 120	**10.** 0·019
11. 19·5	**12.** 1·767	**13.** 30·2	**14.** 0·07	**15.** 1200

Divide each of these numbers by 100.

16. 140	**17.** 32·7	**18.** 0·9	**19.** 2400	**20.** 16·6
21. 3247	**22.** 598	**23.** 4000	**24.** 597·9	**25.** 18·1

$\frac{1}{4}$ can be written as $\frac{25}{100} = 0{\cdot}25$

$\frac{1}{2}$ can be written as $\frac{5}{10} = 0{\cdot}5$

$\frac{3}{4}$ can be written as $\frac{75}{100} = 0{\cdot}75$

Write these as decimals.

26. $1\frac{3}{4}$	**27.** $4\frac{1}{2}$	**28.** $7\frac{1}{4}$	**29.** $8\frac{1}{2}$	**30.** $2\frac{3}{4}$
31. $4\frac{1}{2}$ kg	**32.** $6\frac{1}{2}$ l	**33.** $3\frac{3}{4}$ m	**34.** $2\frac{1}{4}$ kg	**35.** $7\frac{3}{4}$ l
36. $\frac{3}{4}$ kg	**37.** $\frac{1}{2}$ m	**38.** $2\frac{1}{2}$ km	**39.** $1\frac{1}{4}$ l	**40.** $1\frac{3}{4}$ kg
41. $2\frac{1}{2}$ kg	**42.** $\frac{3}{4}$ m	**43.** $\frac{3}{4}$ l	**44.** $5\frac{1}{4}$ kg	**45.** $6\frac{3}{4}$ l

Complete these:

46. 1·5 m = □ cm	**47.** 1·75 kg = □ g	**48.** 0·25 l = □ ml
49. 0·75 m = □ cm	**50.** 2·5 kg = □ g	**51.** 1·75 l = □ ml
52. 3·25 m = □ cm	**53.** 0·25 kg = □ g	**54.** 2·5 l = □ ml
55. 4·5 m = □ cm	**56.** 1·5 kg = □ g	**57.** 0·75 l = □ ml

Write the value of the underlined digit.

58. $1{\cdot}\underline{7}8$	**59.** $\underline{2}3{\cdot}45$	**60.** $0{\cdot}76\underline{3}$	**61.** $1{\cdot}0\underline{8}$	**62.** $0{\cdot}00\underline{2}$
63. $\underline{2}47{\cdot}6$	**64.** $13{\cdot}9\underline{4}$	**65.** $1{\cdot}00\underline{2}$	**66.** $5{\cdot}\underline{7}62$	**67.** $17{\cdot}0\underline{3}$
68. $1\underline{3}1{\cdot}4$	**69.** $246{\cdot}\underline{1}$	**70.** $73{\cdot}9\underline{5}$	**71.** $1{\cdot}\underline{6}03$	**72.** $0{\cdot}4\underline{0}2$

1.
```
    14·7
     3·68
+    1·752
---------

---------
```

2.
```
    6·478
×       4
---------

---------
```

3.
```
  176·5
−  14·75
---------

---------
```

4. 7) 7·329

5.
```
   29·86
×      8
---------

---------
```

6. 9) 8·856

7.
```
    7·238
    1·5
+   0·762
---------

---------
```

8.
```
   15·84
−   9·972
---------

---------
```

9.
```
    1·01
−   0·743
---------

---------
```

10. 6) 16·488

11.
```
   170·6
×      7
---------

---------
```

12.
```
   124·2
    10·75
+  240·0
---------

---------
```

13. Add together 4·74, 16·9 and 1·75.

14. Divide 217·92 by 8.

15. Find the difference between 36·85 and 108.

16. The total of three numbers is 1·748.
Two of the numbers are 0·134 and 0·659.
Find the third number.

17. Find the product of 4 and 0·768.

18. When nine is multiplied by a number, the answer is 34·128.
Find the number.

19. When 4·65 is taken away from a number the answer is 2·493.
Find the number.

20. Multiply 5 by 0·948.

21. The product of two numbers is 81·18. One of the numbers is 9.
What is the other number?

Fill in the missing numbers.

22.
```
   2*6·1
+   18·*
---------
   304·8
---------
```

23.
```
   11·0*
+   3·*9
---------
   14·41
---------
```

24.
```
    4·*6
−   1·3*
---------
    3·66
---------
```

25.
```
    *·35
−   2·**
---------
    6·87
---------
```

Measurement

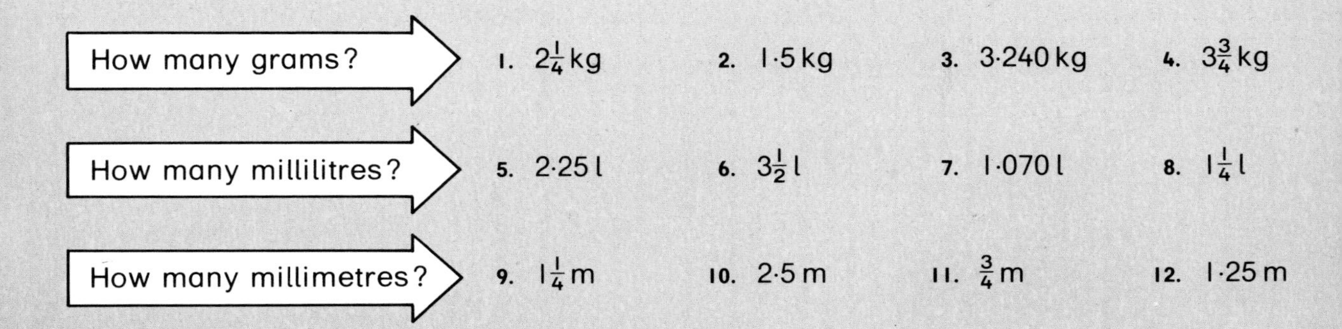

How many grams?

1. $2\frac{1}{4}$ kg 2. 1·5 kg 3. 3·240 kg 4. $3\frac{3}{4}$ kg

How many millilitres?

5. 2·25 l 6. $3\frac{1}{2}$ l 7. 1·070 l 8. $1\frac{1}{4}$ l

How many millimetres?

9. $1\frac{1}{4}$ m 10. 2·5 m 11. $\frac{3}{4}$ m 12. 1·25 m

13. Find the total of B, C and D.
14. Find the difference between A and C.
15. Find a half of C.
16. Double B and write the answer in kg.

A	3·400 kg
B	800 g
C	$1\frac{3}{4}$ kg
D	2·5 kg

17. What is the total capacity of E, F and G?
18. What is the difference between F and H?
19. What is $\frac{3}{5}$ of E?
20. What is three times the capacity of G?

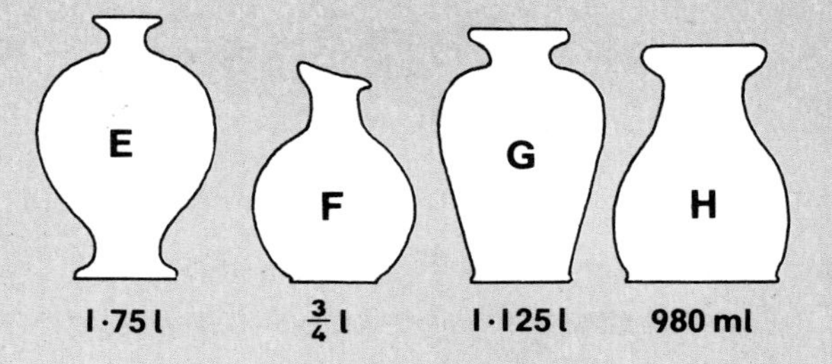

21. How much taller is Mark than Angela?
22. How much shorter is Clive than Myra?
23. When Clive has grown another 30 mm how tall will he be?
24. What is the average height of the four children?

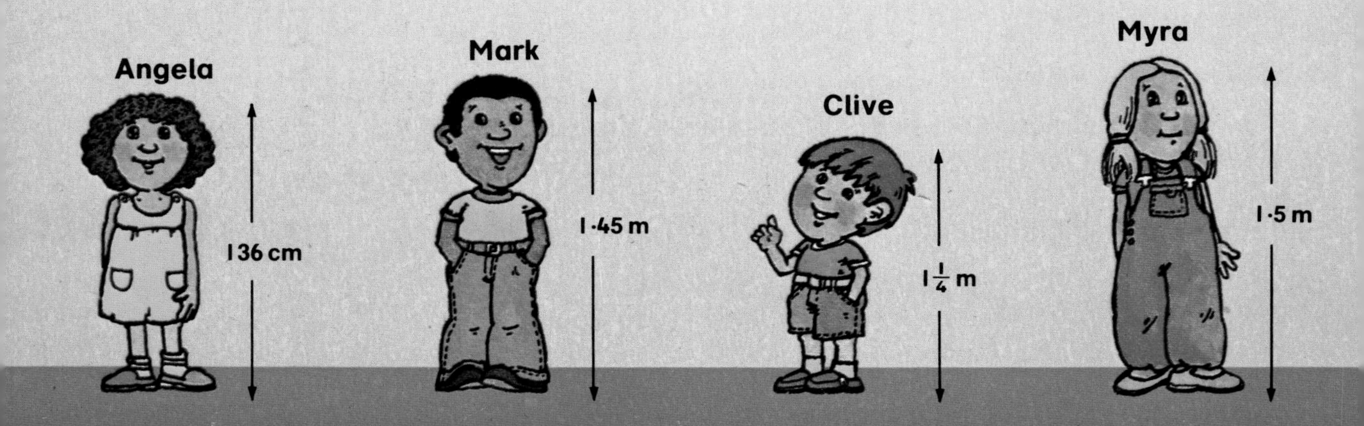

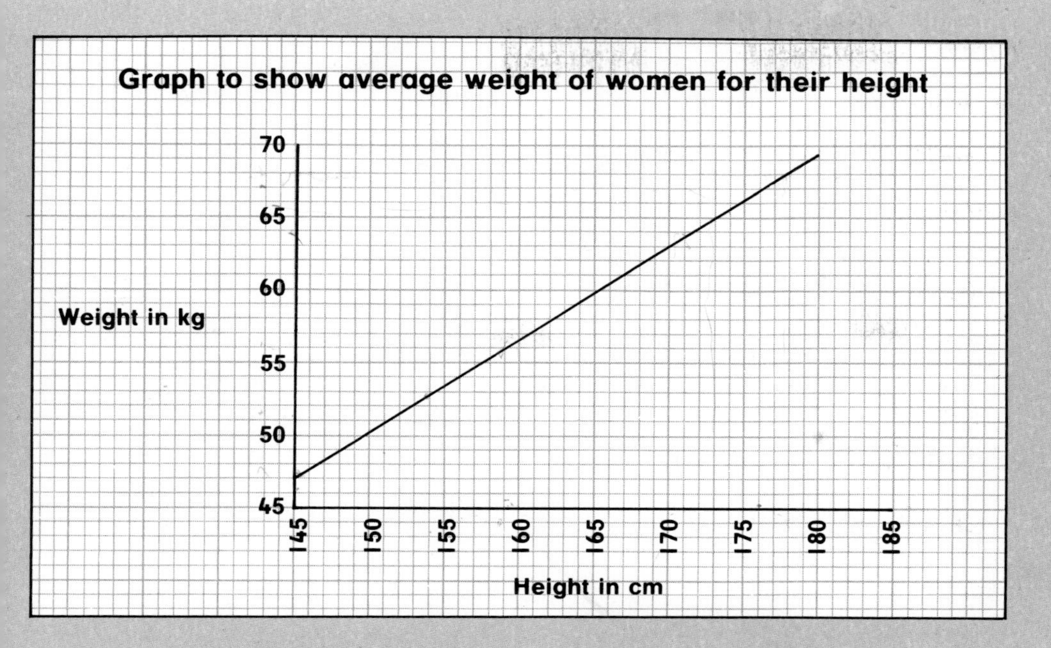

Below are the heights of some women.
What should their weights be to the nearest $\frac{1}{2}$ kg?

1. 156 cm
2. 163 cm
3. 171 cm
4. 174 cm
5. 178 cm

Below are the weights of some women.
What should their heights be to the nearest cm?

6. 47 kg
7. $52\frac{1}{2}$ kg
8. 56 kg
9. 67 kg
10. $68\frac{1}{2}$ kg

Calculate, to the nearest $\frac{1}{2}$ kg, how much these women are above or below the average weight.
Use the graph to help you.

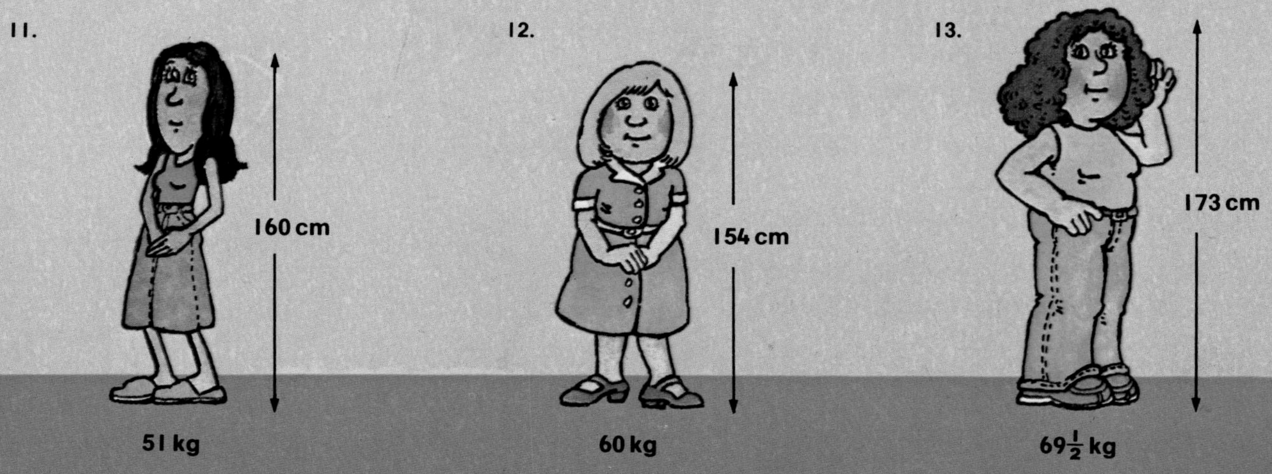

Number

Write the answers only.

1. 49 ÷ 7
2. 56 ÷ 8
3. 65 ÷ 6
4. 53 ÷ 9
5. 44 ÷ 7
6. 50 ÷ 5
7. 42 ÷ 8
8. 19 ÷ 3
9. 27 ÷ 9
10. 58 ÷ 7
11. 33 ÷ 6
12. 51 ÷ 9
13. 39 ÷ 4
14. 13 ÷ 2
15. 63 ÷ 9
16. 74 ÷ 8
17. 47 ÷ 5
18. 59 ÷ 8
19. 55 ÷ 10
20. 40 ÷ 10
21. 37 ÷ 4
22. 73 ÷ 7
23. 25 ÷ 3
24. 84 ÷ 10
25. 33 ÷ 5
26. $7\overline{)6298}$
27. $9\overline{)4372}$
28. $6\overline{)3064}$
29. $8\overline{)7391}$
30. £ $5\overline{)17{\cdot}45}$
31. £ $7\overline{)24{\cdot}36}$
32. £ $10\overline{)19{\cdot}50}$
33. £ $8\overline{)29{\cdot}28}$
34. kg $3\overline{)2{\cdot}916}$
35. kg $10\overline{)4{\cdot}380}$
36. kg $7\overline{)5{\cdot}614}$
37. kg $9\overline{)5{\cdot}787}$
38. l $5\overline{)4{\cdot}985}$
39. l $8\overline{)7{\cdot}392}$
40. l $4\overline{)9{\cdot}364}$
41. l $9\overline{)0{\cdot}297}$

42. A piece of material is $27\frac{1}{2}$ m long.
It is cut into 5 equal pieces.
How long will each piece be?

43. 6 boys weigh a total of 192 kg.
What is their average weight?

44. Father shares £250 equally among his four children.
How much will each get?

45. Eight tins of peaches weigh 3·632 kg.
How much does one tin weigh?

46. Six workmen earn £852 between them each week.
They all earn the same amount.
How much does each one earn?

47. A container of orange juice holds $6\frac{3}{4}$ litres.
It is poured into 9 jugs, all the same size.
There is none left over.
How much does each jug hold?

Which of these are multiples of 10?

4620 3900 2962 1689 4000 2720 599 7030 160 872

Sometimes we need to divide by numbers greater than 10.
To divide by 12, a table of multiples of 12 is a help.

× 1	× 2	× 3	× 4	× 5	× 6	× 7	× 8	× 9
12	24	36	48	60	72	84	96	108

We can use the table to help to work this out.

$$\begin{array}{r} 24 \\ 12\overline{)288} \\ -24 \\ \hline 48 \\ -48 \\ \hline 0 \end{array}$$

Do these in the same way.

1. $12\overline{)276}$
2. $12\overline{)612}$
3. $12\overline{)540}$
4. $12\overline{)408}$
5. $15\overline{)525}$
6. $17\overline{)391}$
7. $14\overline{)588}$
8. $16\overline{)352}$
9. $11\overline{)869}$
10. $20\overline{)540}$
11. $21\overline{)756}$
12. $19\overline{)969}$
13. $25\overline{)1975}$
14. $22\overline{)1826}$
15. $18\overline{)5418}$
16. $26\overline{)2912}$
17. $864 \div 12$
18. $913 \div 11$
19. $864 \div 24$
20. $1292 \div 19$
21. $1955 \div 23$
22. $930 \div 15$
23. $2904 \div 22$
24. $1728 \div 16$
25. The product of two numbers is 1632.
    One number is 12.
    What is the other number?
26. The product of two numbers is 2052.
    One number is 19.
    What is the other number?

Shape

Copy these shapes and descriptions into your book.

Four equal sides

Parallel sides

Opposite angles are equal

Four right angles

Opposite sides are equal

Has rotational symmetry

Map the drawings to the descriptions.

1. How is a square similar to a parallelogram?
2. How is a rhombus different from a square?
3. How is a parallelogram similar to a rhombus?
4. How is a rectangle similar to a parallelogram?
5. How is a square different from a rectangle?
6. Which shape has one line of symmetry?
7. Which shape has no line of symmetry?

A pair of compasses, glue

This is a **tetrahedron**.

This is how you can make one, using a ruler, pencil and compasses.

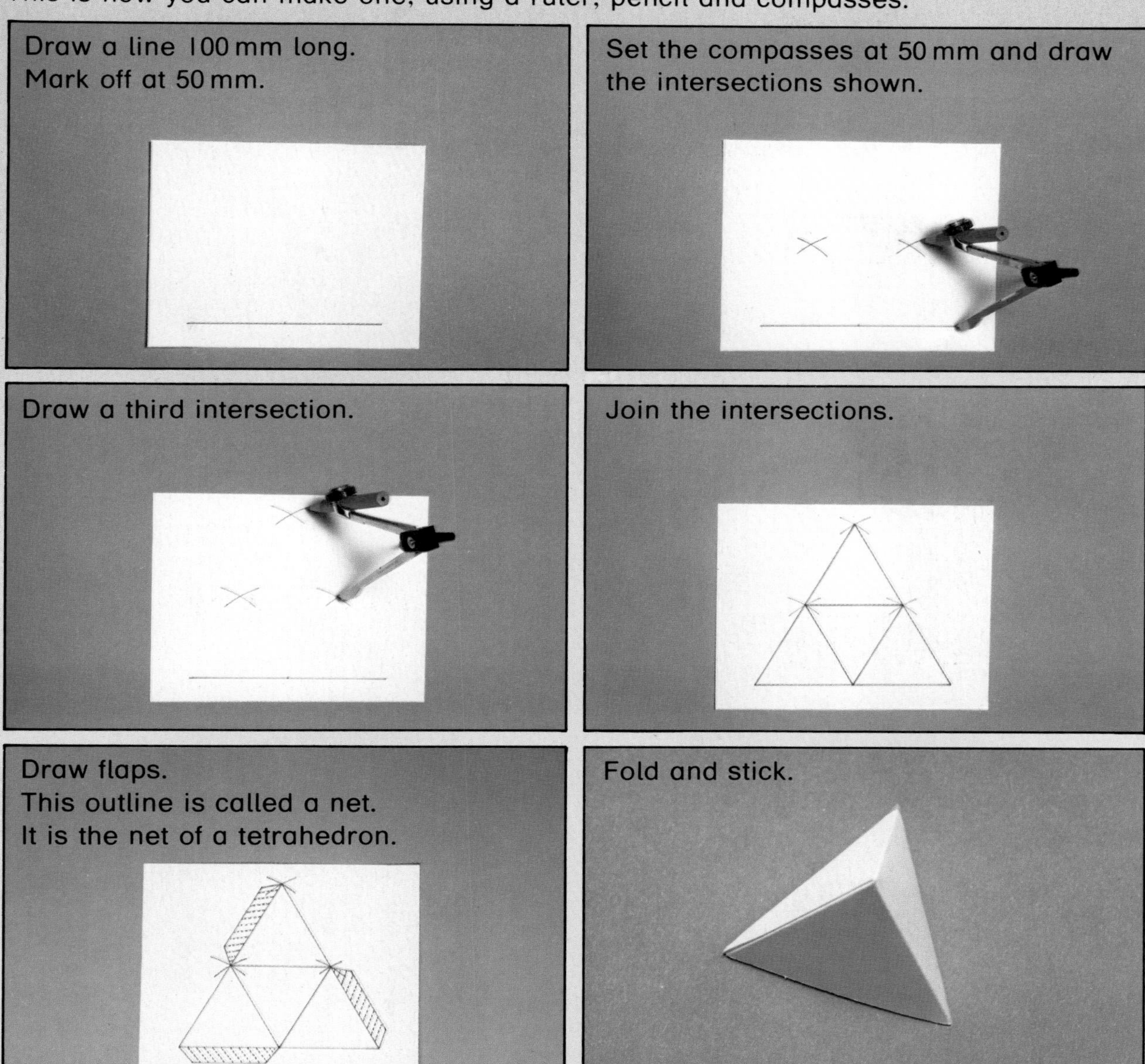

1. How many faces has your tetrahedron?
2. What shape is each face?

Fractions

1. $7\frac{1}{4} + 3\frac{2}{3}$
2. $1\frac{7}{8} + 4\frac{1}{3}$
3. $4\frac{3}{5} + 2\frac{5}{7}$
4. $6\frac{4}{5} + 2\frac{1}{4}$
5. $\frac{7}{9} + 2\frac{5}{6}$
6. $3\frac{5}{8} + 1\frac{4}{5}$
7. $2\frac{1}{3} + 5\frac{4}{5}$
8. $4\frac{8}{9} + 1\frac{2}{3}$
9. $1\frac{2}{7} + 3\frac{1}{2}$
10. $4\frac{3}{10} + 1\frac{2}{3}$
11. $\frac{7}{8} + \frac{9}{10}$
12. $2\frac{5}{8} + 3\frac{3}{5}$
13. $7\frac{1}{3} - 1\frac{3}{4}$
14. $4\frac{1}{9} - 2\frac{1}{4}$
15. $6\frac{1}{4} - \frac{5}{6}$
16. $2\frac{3}{8} - 1\frac{1}{2}$
17. $5\frac{2}{3} - 1\frac{6}{7}$
18. $3\frac{1}{10} - 2\frac{2}{3}$
19. $8\frac{3}{5} - 3\frac{3}{4}$
20. $2\frac{1}{12} - \frac{3}{4}$
21. $4\frac{2}{5} - 1\frac{3}{4}$
22. $5 - 2\frac{4}{9}$
23. $6\frac{1}{4} - 4\frac{5}{6}$
24. $2\frac{4}{9} - \frac{7}{9}$

When multiplying fractions, multiply the numerators together, and then multiply the denominators together.

Write the answer in its simplest form.

$$\frac{2}{7} \times \frac{1}{2}$$
$$= \frac{2}{14}$$
$$= \frac{1}{7}$$

$$\frac{1}{2} \times 2 = 1$$
$$\frac{1}{2} \times 1 = \frac{1}{2}$$
$$\frac{1}{2} \times \frac{1}{2} = \frac{1}{4}$$

25. $\frac{1}{4} \times \frac{1}{2}$
26. $\frac{1}{5} \times \frac{1}{2}$
27. $\frac{1}{3} \times \frac{1}{2}$
28. $\frac{2}{3} \times \frac{1}{2}$
29. $\frac{2}{5} \times \frac{1}{3}$
30. $\frac{1}{2} \times \frac{2}{5}$
31. $\frac{3}{4} \times \frac{4}{9}$
32. $\frac{1}{2} \times \frac{2}{7}$
33. $\frac{1}{12} \times \frac{3}{4}$
34. $\frac{2}{9} \times \frac{3}{4}$
35. $\frac{5}{6} \times \frac{3}{10}$
36. $\frac{7}{8} \times \frac{2}{7}$
37. $\frac{9}{10} \times \frac{5}{6}$
38. $\frac{3}{8} \times \frac{2}{9}$
39. $\frac{7}{10} \times \frac{2}{5}$
40. $\frac{2}{3} \times \frac{9}{10}$

When multiplying fractions, it is sometimes possible to reduce a numerator and denominator before multiplying.

$$\frac{\overset{1}{\cancel{3}}}{4} \times \frac{1}{\underset{2}{\cancel{6}}}$$

$$= \frac{1}{8}$$

1. $\frac{2}{5} \times \frac{1}{2}$
2. $\frac{4}{9} \times \frac{3}{4}$
3. $\frac{3}{4} \times \frac{1}{12}$
4. $\frac{2}{7} \times \frac{1}{2}$
5. $\frac{3}{4} \times \frac{2}{9}$
6. $\frac{2}{7} \times \frac{7}{8}$
7. $\frac{3}{10} \times \frac{5}{6}$
8. $\frac{5}{6} \times \frac{9}{10}$
9. $\frac{2}{5} \times \frac{7}{10}$
10. $\frac{2}{9} \times \frac{3}{8}$
11. $\frac{2}{3} \times \frac{7}{10}$
12. $\frac{2}{5} \times \frac{1}{4}$
13. $\frac{11}{12} \times \frac{3}{5}$
14. $\frac{4}{9} \times \frac{3}{10}$
15. $\frac{5}{16} \times \frac{4}{5}$
16. $\frac{11}{12} \times \frac{4}{7}$
17. $\frac{7}{16} \times \frac{6}{7}$
18. $\frac{8}{9} \times \frac{3}{16}$
19. $\frac{5}{6} \times \frac{4}{9}$
20. $\frac{2}{3} \times \frac{1}{2}$
21. $\frac{3}{4} \times \frac{8}{9}$
22. $\frac{3}{4} \times \frac{5}{12}$
23. $\frac{4}{7} \times \frac{1}{2}$
24. $\frac{4}{7} \times \frac{7}{8}$
25. $\frac{1}{10} \times \frac{5}{6}$
26. $\frac{4}{5} \times \frac{7}{10}$
27. $\frac{4}{9} \times \frac{3}{8}$
28. $\frac{2}{3} \times \frac{9}{10}$
29. $\frac{2}{5} \times \frac{3}{4}$
30. $\frac{5}{12} \times \frac{4}{5}$
31. $\frac{3}{4} \times \frac{4}{9}$
32. $\frac{1}{3} \times \frac{9}{10}$

33. There are 320 books on a shelf.
$\frac{3}{4}$ of the books are fiction.
How many of the books are non-fiction?

34. A box holds 240 beads.
$\frac{1}{6}$ of the beads are red.
$\frac{1}{8}$ of these red beads are square.
How many square red beads are there?

35. In a school there are 250 children.
$\frac{3}{5}$ of them are boys.
$\frac{1}{3}$ of the boys are under eight years old.
How many boys are under eight years old?

Number

This table shows the money taken by a supermarket in the last 4 weeks.

Week No. 1	Week No. 2	Week No. 3	Week No. 4
£16218	£12936	£13626	£14616

1. Find the difference in money taken in weeks 1 and 2.
2. Find the difference in money taken in weeks 3 and 4.
3. Find the total amount of money taken in the 4 weeks.
4. Find the average amount of money taken each week.
5. $\frac{1}{3}$ of the money taken is profit.
   What was the total profit for the 4 weeks?
6. $\frac{1}{12}$ of the money taken was spent on heating and lighting.
   How much did it cost?
7. The supermarket is open 6 days a week.
   Find the average daily takings for

   a) week 1. b) week 2. c) week 3. d) week 4.

1. How much weight on each of the 4 shelves?
2. Work out the value of each shelf of items.

Measurement

Stop-watch, pendulum

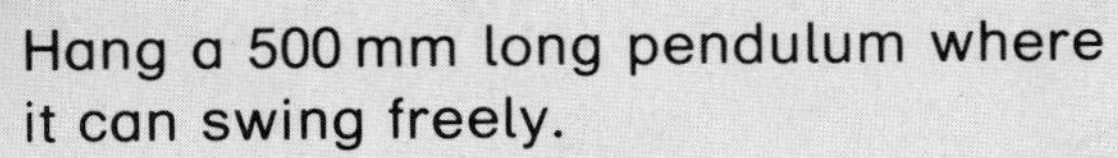

Hang a 500 mm long pendulum where it can swing freely.

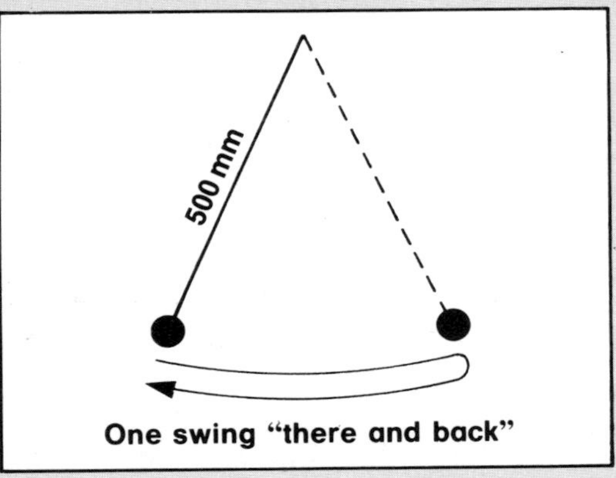

One swing "there and back"

One swing is when the pendulum returns to where it started.

Start the pendulum swinging.
Time 5 swings.
Keep the pendulum the same length.
Time 5 longer swings.
Time 5 shorter swings.
What do you notice?

Time 5 swings for pendulums of length:
400 mm; 300 mm; 200 mm and 100 mm.
Record your results in a table.

Length of pendulum in mm	500	400	300	200	100
Time for 5 swings					

How does the length of a pendulum alter the time it takes to swing?

Shape

Draw a 60 mm circle.
Construct a hexagon in the circle.

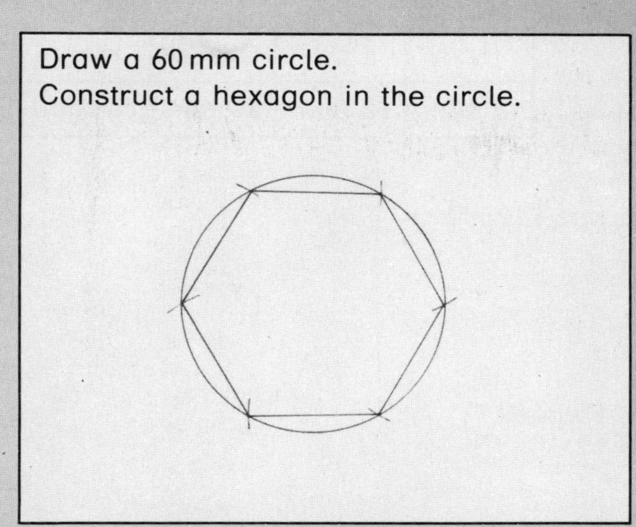

Extend some of the sides of the hexagon to make these four equilateral triangles.

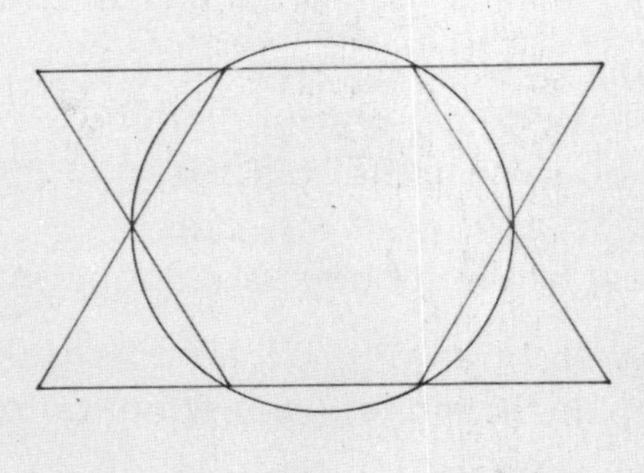

Construct two squares on opposite sides of the hexagon.
Add two flaps to your net as shown.

Cut out the net and stick the shaded parts.

Make another identical solid.
Now put your two shapes together to make a tetrahedron.

Fractions

This table shows the time taken to travel 5 km.

Train	Car	Cyclist	Hiker
3 min.	12 min.	20 min.	50 min.

1. Write each time as a fraction of an hour.
2. How far would each travel in one hour?

The cost of building an extension to a house was £12500.
The bricks cost $\frac{1}{5}$ of that amount.
Other building materials cost $\frac{2}{5}$ of the total.
Transport cost $\frac{1}{10}$ of the total.
The rest of the money was spent on wages.

3. What fraction of the total cost was spent on wages?
4. How much money was spent on each item?

Mr. White is a milkman.
He calls at 240 houses a day.
At $\frac{1}{3}$ of his calls he delivers 1 bottle of milk.
At $\frac{1}{2}$ of his calls he delivers 2 bottles of milk.
At the rest of his calls he delivers 3 bottles of milk.

5. How many bottles of milk does he deliver each day?

In a tennis tournament the total prize money was £600.
The prizes were: 1st prize £300
2nd prize £150
3rd prize £100
4th prize £50

6. Write each prize as a fraction of the total prize money.

Before multiplying mixed numbers, change them to fractions.

$1\frac{1}{4} \times \frac{2}{5}$	$\frac{3}{5} \times 2\frac{1}{2}$
$= \frac{\cancel{5}^{1}}{\cancel{4}_{2}} \times \frac{\cancel{2}^{1}}{\cancel{5}_{1}}$	$= \frac{3}{\cancel{5}_{1}} \times \frac{\cancel{5}^{1}}{2}$
$= \frac{1}{2}$	$= \frac{3}{2}$
	$= 1\frac{1}{2}$

1. $2\frac{2}{3} \times \frac{9}{16}$
2. $\frac{4}{7} \times 1\frac{3}{4}$
3. $1\frac{3}{8} \times \frac{4}{11}$
4. $\frac{6}{7} \times 1\frac{1}{6}$
5. $\frac{2}{7} \times 3\frac{1}{2}$
6. $2\frac{2}{5} \times \frac{5}{6}$
7. $1\frac{4}{9} \times \frac{3}{5}$
8. $\frac{4}{7} \times 2\frac{5}{8}$
9. $\frac{9}{11} \times 3\frac{2}{3}$
10. $2\frac{3}{4} \times \frac{7}{11}$
11. $\frac{3}{4} \times 1\frac{5}{7}$
12. $4\frac{2}{3} \times \frac{5}{7}$

$1\frac{1}{2} \times 1\frac{1}{3}$

$= \frac{\cancel{3}^{1}}{\cancel{2}_{1}} \times \frac{\cancel{4}^{2}}{\cancel{3}_{1}}$

$= \frac{2}{1}$

$= 2$

13. $1\frac{7}{8} \times 1\frac{3}{5}$
14. $1\frac{1}{9} \times 1\frac{8}{10}$
15. $2\frac{2}{3} \times 1\frac{1}{2}$
16. $1\frac{7}{8} \times 2\frac{2}{5}$
17. $2\frac{2}{3} \times 2\frac{1}{4}$
18. $2\frac{1}{7} \times 4\frac{2}{3}$
19. $1\frac{1}{8} \times 1\frac{1}{3}$
20. $2\frac{1}{2} \times 1\frac{3}{5}$
21. $1\frac{3}{4} \times 1\frac{3}{7}$
22. $2\frac{1}{4} \times 1\frac{1}{6}$
23. $2\frac{7}{8} \times 1\frac{1}{3}$
24. $1\frac{4}{5} \times 1\frac{1}{4}$
25. $1\frac{5}{6} \times 1\frac{1}{8}$
26. $2\frac{2}{5} \times 1\frac{1}{4}$
27. $3\frac{1}{2} \times 1\frac{1}{5}$
28. $1\frac{2}{3} \times 2\frac{4}{5}$

Decimals

Numbers less than 0·5 are rounded down.
Numbers 0·5 and above are rounded up.

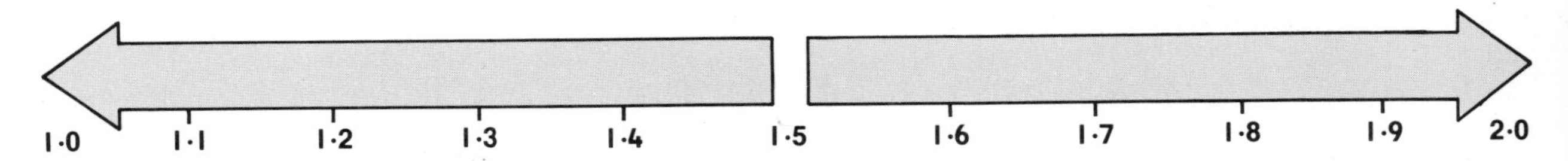

1·1 rounded off to the nearest whole number is 1.
1·8 rounded off to the nearest whole number is 2.
1·5 rounded off to the nearest whole number is 2.

1·75 rounded off to the nearest whole number is 2.
This is an approximation.
It can be written like this: 1·75 ≏ 2

≏ means **approximately equal to.**

Round these numbers off to the nearest whole number, using the approximation sign.

1. 1·2	2. 1·6	3. 1·9	4. 1·4	5. 1·25
6. 1·35	7. 1·57	8. 1·02	9. 1·76	10. 1·69
11. 1·452	12. 1·917	13. 1·502	14. 1·999	15. 1·499
16. 3·7	17. 8·1	18. 5·5	19. 7·8	20. 2·1
21. 0·8	22. 1·74	23. 6·95	24. 4·592	25. 8·455
26. 11·8	27. 0·946	28. 7·601	29. 5·511	30. 6·099

Which of these numbers are between 8 and 9?

31. 6·218, 8·71, 7·3, 8·102, 9·029

Which of these numbers are between 60 and 70?

32. 56·421, 62·02, 69·115, 71·2, 70·2

Which of these numbers are between 120 and 130?

33. 112·21, 126·205, 130·8, 129·2, 119·08

These numbers are written to **1 decimal place**. 3·8, 42·7, 0·9

Work out these to 1 decimal place.

1. 6·8 ÷ 4
2. 3·5 ÷ 7
3. 29·2 ÷ 4
4. 21·8 ÷ 2
5. 67·1 ÷ 11
6. 50·4 ÷ 12
7. 100·8 ÷ 9
8. 27·3 ÷ 13

13 ÷ 2 can be worked out to one decimal place, like this: $2\overline{)13\cdot0}$ with answer 6·5

Noughts added to a whole number, after the decimal point, do not alter its value.
Do these in the same way.

9. 17 ÷ 2
10. 18 ÷ 4
11. 26 ÷ 5
12. 20 ÷ 8

These numbers are written to **2 decimal places**. 5·68, 0·74, 54·07

Work out these to two decimal places.

13. 17·32 ÷ 4
14. 6·51 ÷ 7
15. 13·35 ÷ 5
16. 53·28 ÷ 8
17. 36·99 ÷ 3
18. 43·68 ÷ 6
19. 15·68 ÷ 14
20. 33·15 ÷ 15
21. 17 ÷ 4
22. 13·2 ÷ 8
23. 13·7 ÷ 5
24. 10 ÷ 8

These numbers are written to **3 decimal places**. 7·962, 0·073, 1·107

Work out these to 3 decimal places.

25. 1·655 ÷ 5
26. 8·694 ÷ 7
27. 16·362 ÷ 3
28. 2·808 ÷ 8
29. 3·915 ÷ 9
30. 16·687 ÷ 11
31. 11 ÷ 8
32. 7·89 ÷ 6

The arrow points to 0·3.

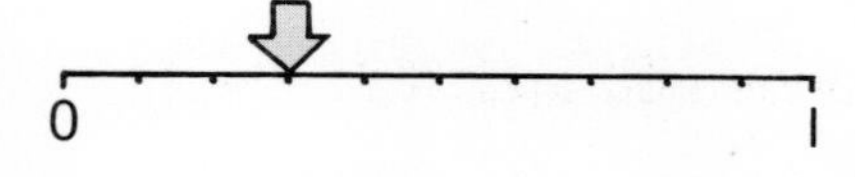

Write the number each arrow points to.

33.

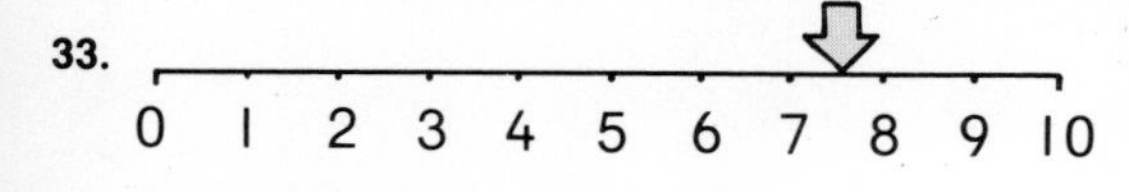

34.

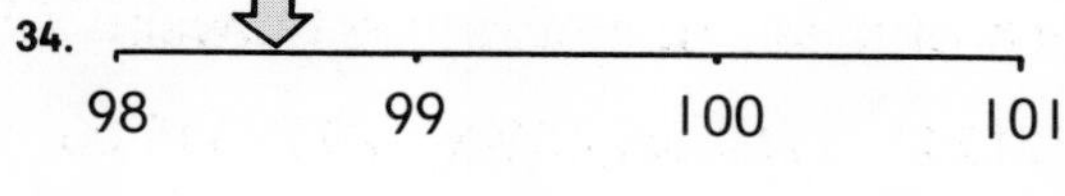

35.

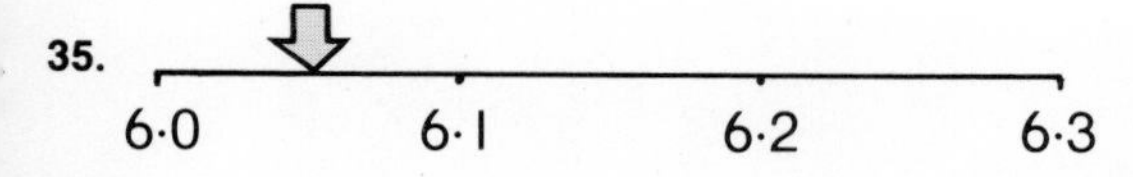

36.

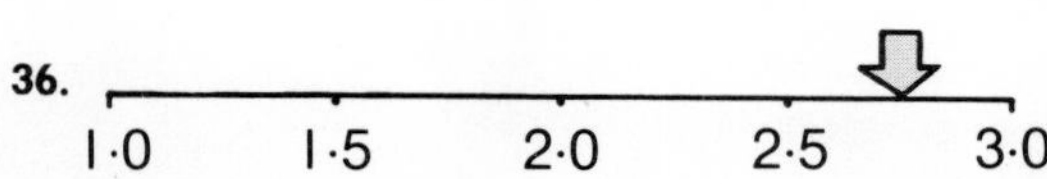

Fractions

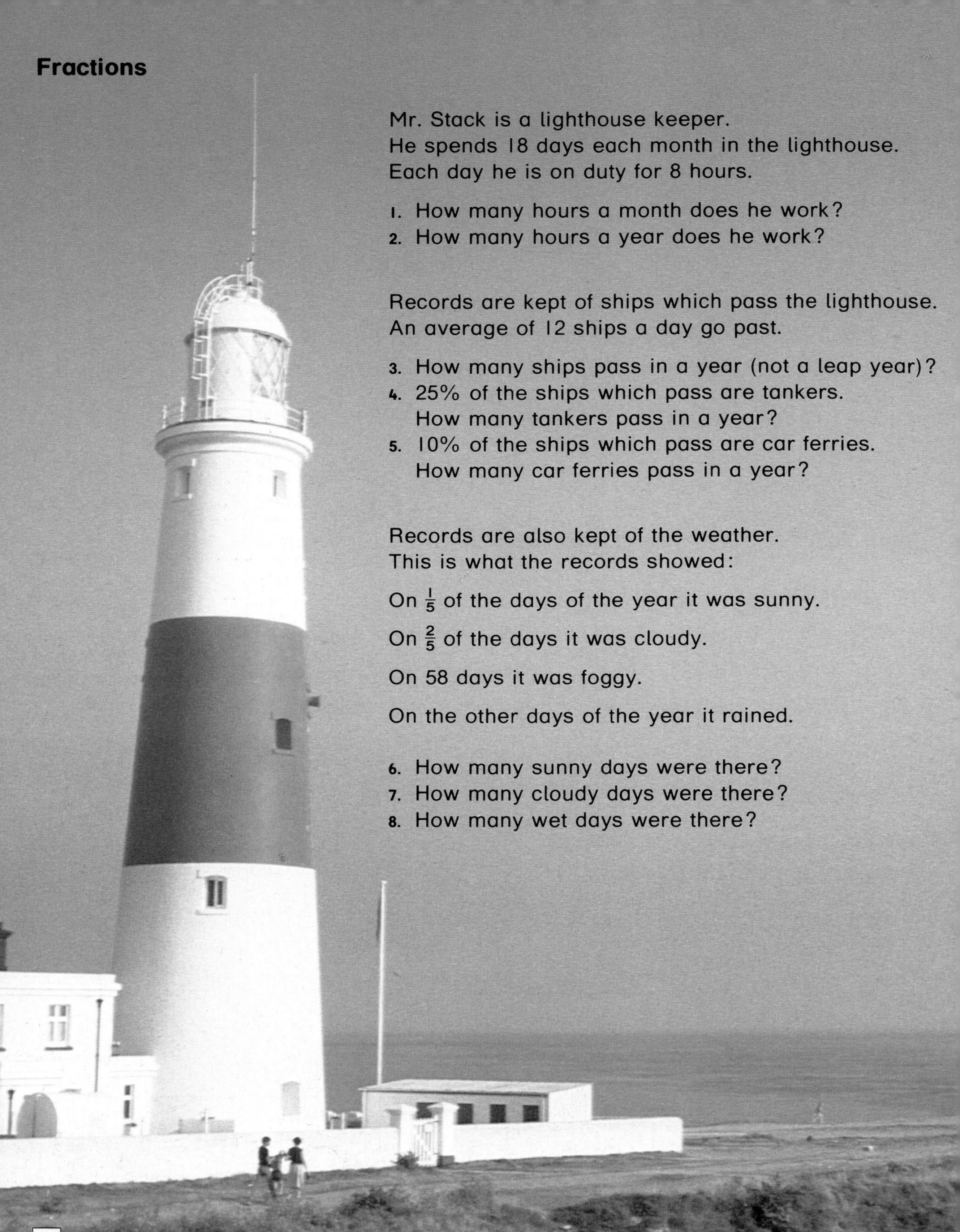

Mr. Stack is a lighthouse keeper.
He spends 18 days each month in the lighthouse.
Each day he is on duty for 8 hours.

1. How many hours a month does he work?
2. How many hours a year does he work?

Records are kept of ships which pass the lighthouse.
An average of 12 ships a day go past.

3. How many ships pass in a year (not a leap year)?
4. 25% of the ships which pass are tankers.
   How many tankers pass in a year?
5. 10% of the ships which pass are car ferries.
   How many car ferries pass in a year?

Records are also kept of the weather.
This is what the records showed:

On $\frac{1}{5}$ of the days of the year it was sunny.

On $\frac{2}{5}$ of the days it was cloudy.

On 58 days it was foggy.

On the other days of the year it rained.

6. How many sunny days were there?
7. How many cloudy days were there?
8. How many wet days were there?

Number

Hundred thousands	Ten thousands	Thousands	Hundreds	Tens	Units

Round off these numbers to the nearest thousand.

1. 1730
2. 4610
3. 5132
4. 11 600
5. 9750
6. 14 146
7. 23 200
8. 37 394
9. 54 725
10. 124 600
11. 138 410
12. 165 460

Write these numbers in order of size, smallest first.

13. 4016, 2995, 13 217, 10 408, 9624
14. 3700, 2850, 16 000, 21 000, 9500
15. 140 300, 138 500, 96 000, 100 000

This table shows the approximate number of people using a port during the seasons of a year.

Season	Number of people
Spring	286 000
Summer	347 000
Autumn	310 000
Winter	164 000

16. In which season was the port busiest?
17. In which season was the port least busy?
18. How many more people used the port in Summer than in Spring?
19. How many fewer people used the port in Winter than in Autumn?
20. What was the total number of people using the port during the year?

You will see that the total number has seven digits in it.
The first digit stands for **millions**.

Millions	Hundred thousands	Ten thousands	Thousands	Hundreds	Tens	Units

21. Write the total number of people using the port in words.

Graphs

These two graphs show the same information.
They show the heights of five teachers at John's school.

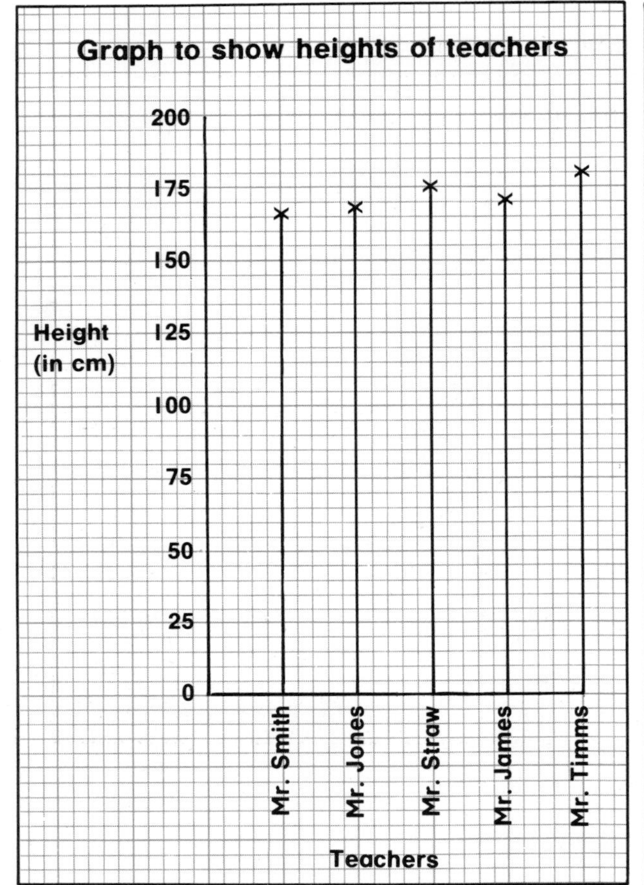

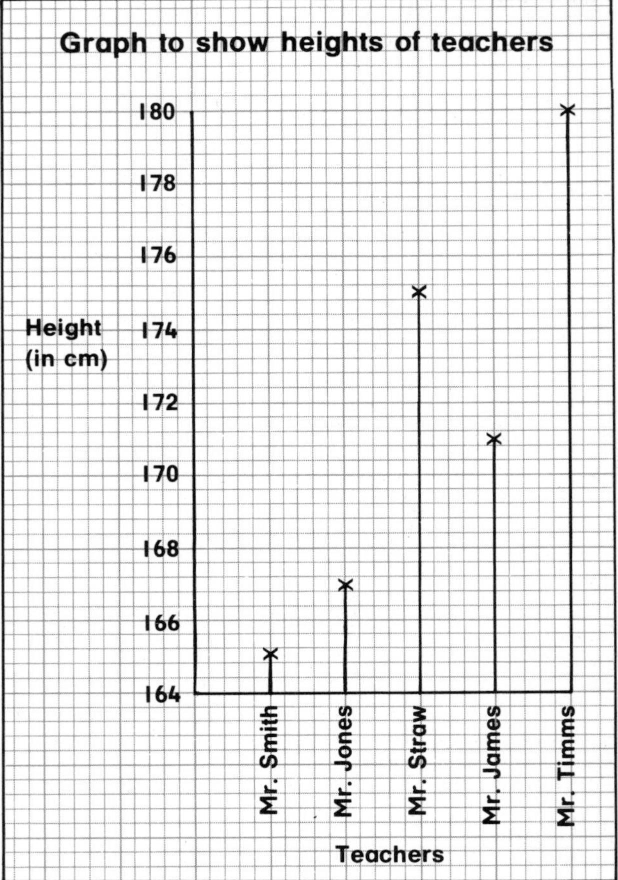

The second graph is easier to read.
The scale on the vertical axis has been changed.
It starts at 164 instead of 0.
It is not always sensible to start the vertical axis at 0.
From the second graph find the heights of the five teachers.

1. How much taller than Mr. Straw is Mr. Timms?
2. How much shorter than Mr. Jones is Mr. Smith?
3. A fence round the school is 1·7 m high. Which teachers are taller than the fence?

This train is travelling from Manchester to London.
The journey is 300 km.
The train travels 60 km each hour.
This graph shows its distance from London after each hour of travel.

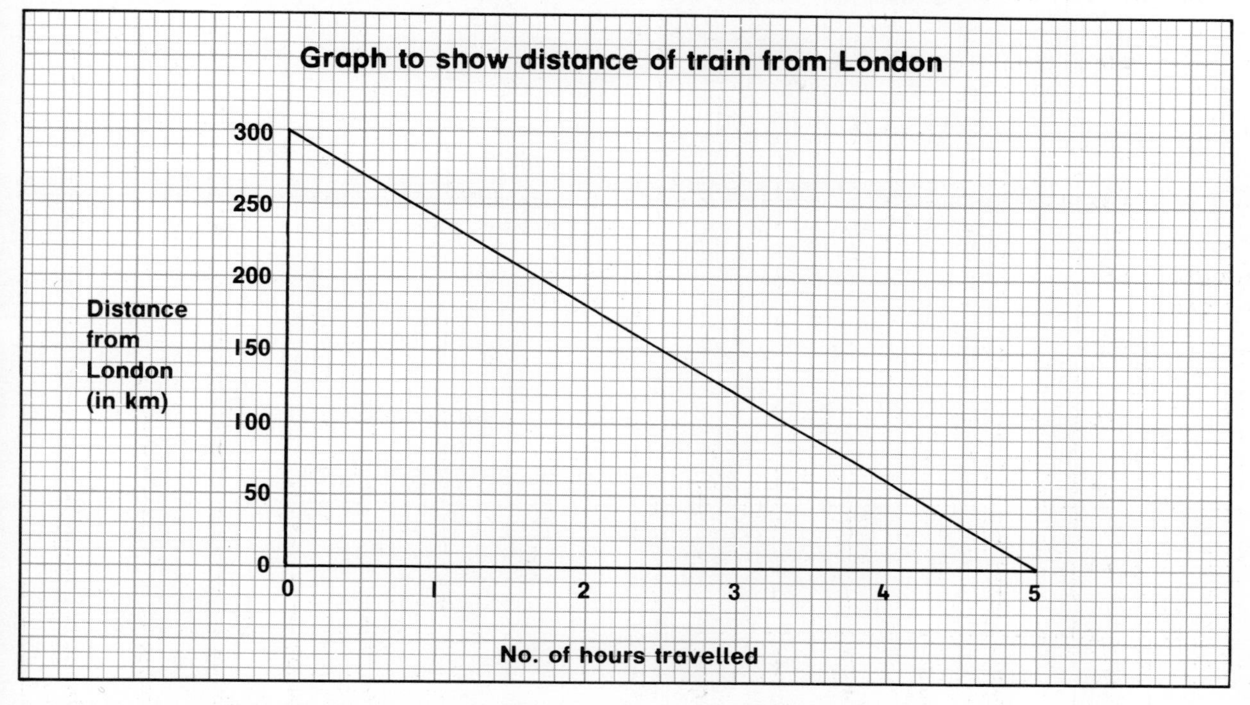

From the graph find the train's distance from London after

1. $1\frac{1}{2}$ hrs; 2. $2\frac{1}{2}$ hrs; 3. $3\frac{1}{2}$ hrs; 4. $4\frac{1}{2}$ hrs.

5. Draw another graph to show the distance from London for a train doing the same journey at 50 km each hour.

Shape

This is a triangular prism.

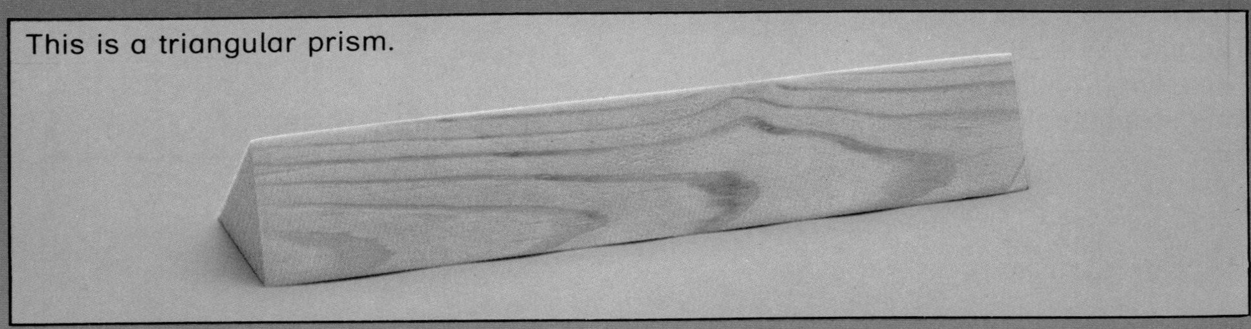

It is the same shape all the way through.

Prisms are the same shape all the way through.

Which of these shapes are prisms?

Area

Greywall is a small village.
It has just had a play park made for the children.
This plan shows the layout of it.
It is drawn to the scale 1 cm : 10 m.

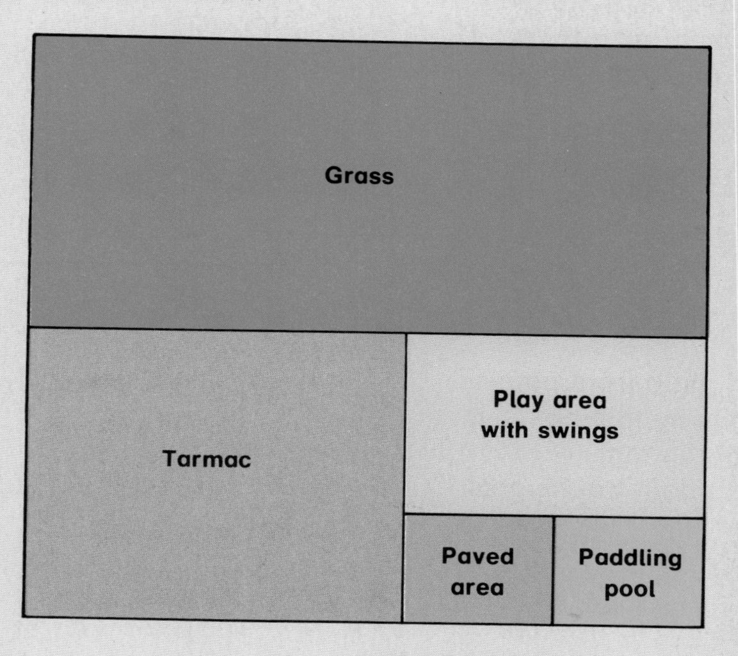

The whole park

1. What is its area?
2. The land cost £2 a square metre. How much did the park cost?

The play area

3. What is its area?
4. It has 6 sets of swings at £68·50 a set. How much did the swings cost?

The tarmac

5. What is its area?
6. Tarmac costs £2·50 a m^2. How much did it cost to cover this area with tarmac?

The grassy part

7. What is its area?
8. Turf costs £1 for 10 m^2. How much did it cost to cover this part with turf?

The paddling pool

9. What is its area?
10. It is $\frac{1}{2}$ metre deep. What is its volume?

The paved area

11. What is its area?
12. It is paved with slabs 1 m by $\frac{1}{2}$ m. How many slabs are there?

The perimeter of the park

13. What is the perimeter of the park?
14. What is the perimeter of each separate area?

Measurement

Cut a scalene triangle from paper.
Ask your teacher to show you how to find a height by folding.
Find the other two heights by folding.
Measure the three heights in mm.
What do you notice about the heights crossing each other?

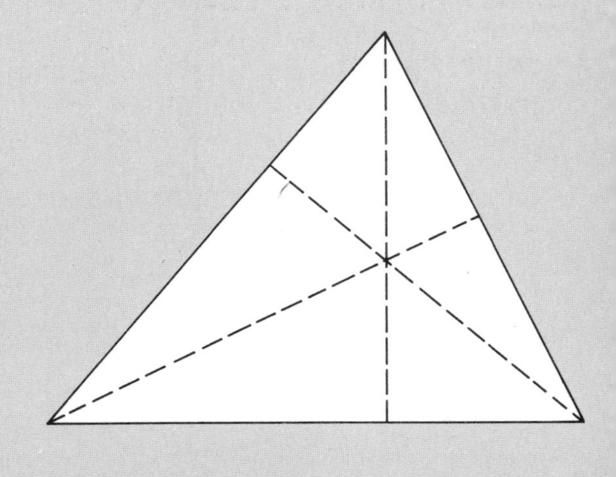

Cut an isosceles triangle from paper.
Find its three heights by folding.
Measure each height in mm.
What do you notice about the heights crossing each other?

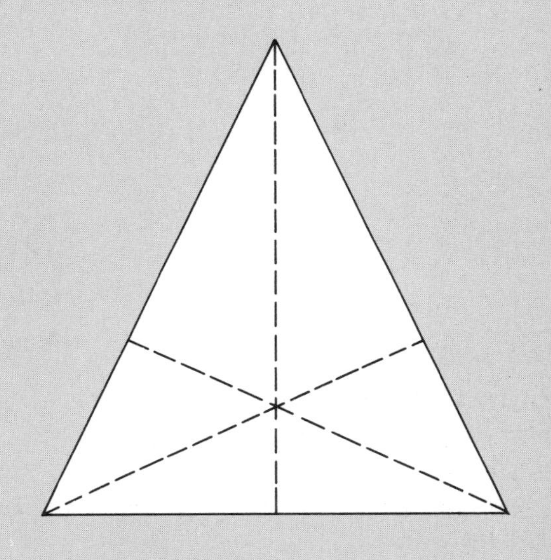

Cut an equilateral triangle from paper.
Find its three heights by folding.
Measure each height in mm.
What do you notice about the heights crossing each other?

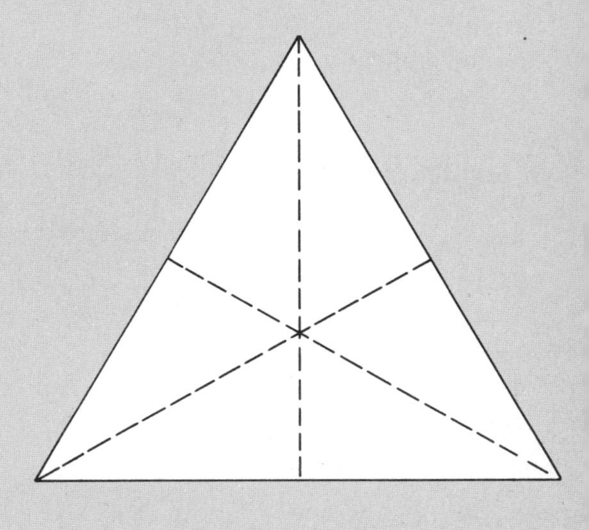

Set square, compasses, protractor

Measure the heights drawn in each of these triangles.

Construct the following triangles.
Use a set square to draw the shortest height in each triangle.
Measure each height you have drawn.

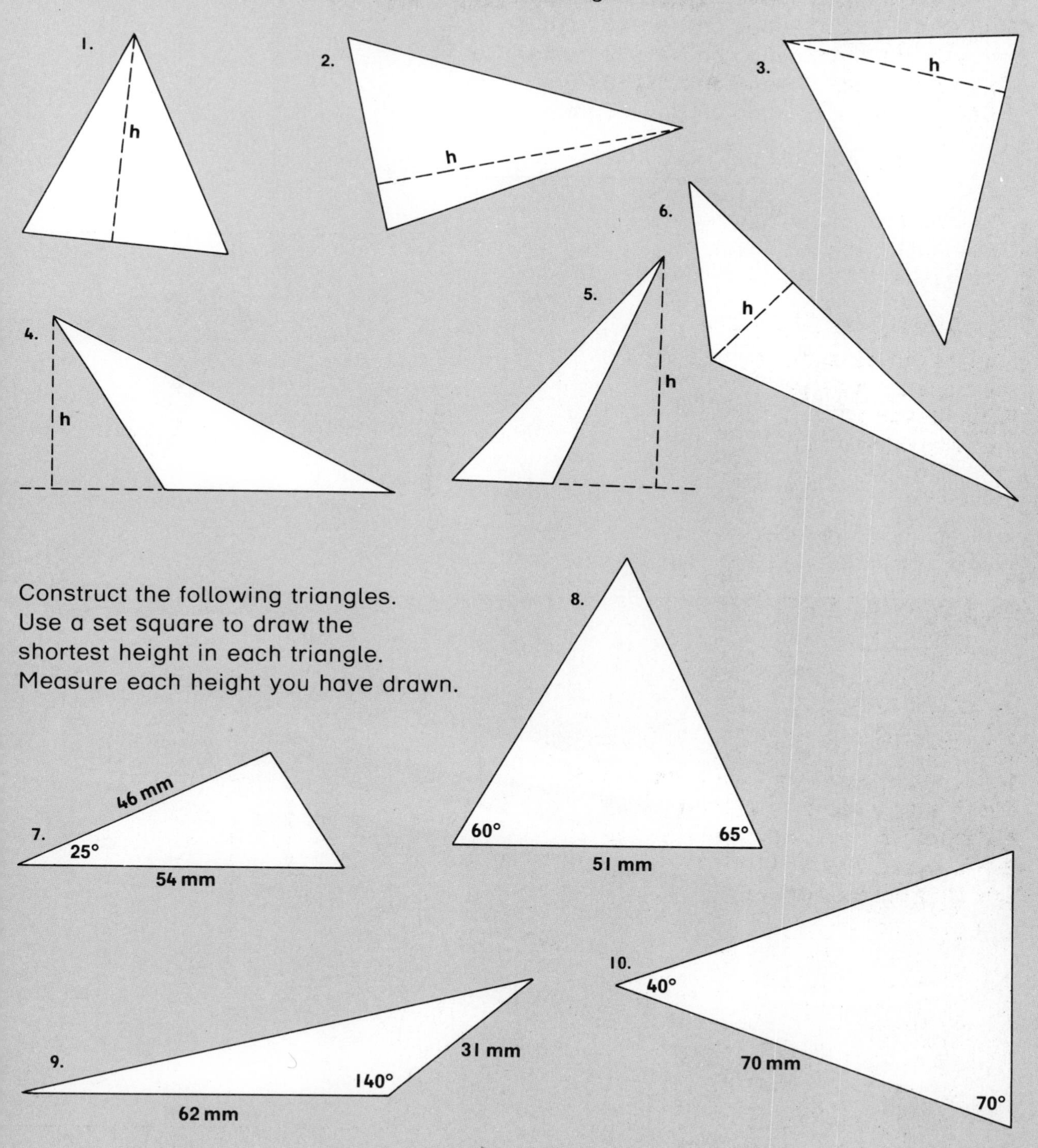

Number

At Crich there is a Tramway Museum.
Some of the trams carry passengers along a tramway.

Seating on No. 180 tram	– 38
Seating on No. 32 tram	– 78
Seating on No. 68 tram	– 56
Seating on No. 166 tram	– 36
Cost per ride – 10p	

No. 180 tram made 2 journeys.
On the first journey it was $\frac{1}{2}$ full.
On the second journey 16 seats were empty.

1. How many people travelled on the tram?
2. How much money was collected in fares?

No. 32 tram made 3 journeys.
On the first journey it was full.
On the second journey there were 4 empty seats.
On the third journey there were 59 passengers.

3. How many people travelled on the tram?
4. How much money was collected in fares?

No. 68 tram made 2 journeys.
On the first journey it was only $\frac{1}{4}$ full.
On the next journey the seats were full, and 8 people were standing.

5. How many people travelled on the tram?
6. How much money was collected in fares?

No. 166 tram made 4 journeys.
On two of the journeys it was $\frac{2}{3}$ full.
On the other two journeys it was $\frac{3}{4}$ full.

7. How many people travelled on the tram?
8. How much money was collected in fares?

9. How many more people travelled on No. 32 tram than on No. 166?
10. How much more was collected on No. 68 tram than on No. 180?
11. How many people travelled on the trams altogether?
12. How much money was collected altogether?

Fractions

The **inverse** of 4 is $\frac{1}{4}$.

An inverse is made by the numerator changing position with the denominator.

The inverse of 8 is $\frac{1}{8}$.

The inverse of $\frac{1}{9}$ is 9.

Write the inverse of these numbers.

1. 7
2. 5
3. $\frac{1}{5}$
4. $\frac{1}{7}$
5. 12
6. $\frac{3}{5}$
7. $\frac{3}{4}$
8. 3
9. 6
10. $\frac{2}{3}$

Dividing by a number is the same as multiplying by its inverse.

$$\frac{2}{3} \div 2$$

$$= \frac{\cancel{2}^{1}}{3} \times \frac{1}{\cancel{2}_{1}}$$

$$= \frac{1}{3}$$

Find the answer to these by multiplying by the inverse.

11. $\frac{2}{3} \div 4$
12. $\frac{2}{7} \div 2$
13. $\frac{5}{6} \div 10$
14. $\frac{3}{4} \div 9$
15. $\frac{4}{5} \div 8$
16. $\frac{5}{8} \div 10$
17. $\frac{2}{3} \div 6$
18. $\frac{2}{7} \div 8$

Multiplying by a number is the same as dividing by its inverse.

$$12 \times \frac{1}{3}$$

$$= 12 \div 3$$

$$= 4$$

Find the answers to these by dividing by the inverse.

19. $16 \times \frac{1}{4}$
20. $14 \times \frac{1}{7}$
21. $48 \times \frac{1}{8}$
22. $30 \times \frac{1}{5}$
23. $18 \times \frac{1}{3}$
24. $25 \times \frac{1}{5}$
25. $45 \times \frac{1}{9}$
26. $36 \times \frac{1}{4}$

Work these out in the easiest way for you.

1. $9 \times \frac{1}{3}$	2. $32 \times \frac{1}{8}$	3. $20 \div 4$	4. $\frac{5}{9} \div 10$	5. $27 \times \frac{1}{9}$
6. $\frac{12}{16} \div 6$	7. $28 \times \frac{1}{4}$	8. 14×7	9. $64 \times \frac{1}{8}$	10. $35 \times \frac{1}{5}$
11. $\frac{7}{8} \div 14$	12. $\frac{5}{6} \div 15$	13. $24 \times \frac{1}{6}$	14. $\frac{4}{7} \div 8$	15. $36 \times \frac{1}{9}$

16. $\frac{2}{3}$ of an apple pie has to be shared equally among 8 people.
What fraction of the whole pie will each person receive?

17. Two girls were given $\frac{3}{4}$ of a cake to share equally between them.
What fraction of the whole cake did each girl receive?

18. Four friends had $\frac{3}{4}$ of a bar of chocolate to share equally amongst themselves.
What fraction of the whole bar did each receive?

19. A bottle of pop was $\frac{3}{5}$ full.
Six boys decided to share it equally.
What fraction of the full bottle did each boy receive?

Area

Calculate the red areas.

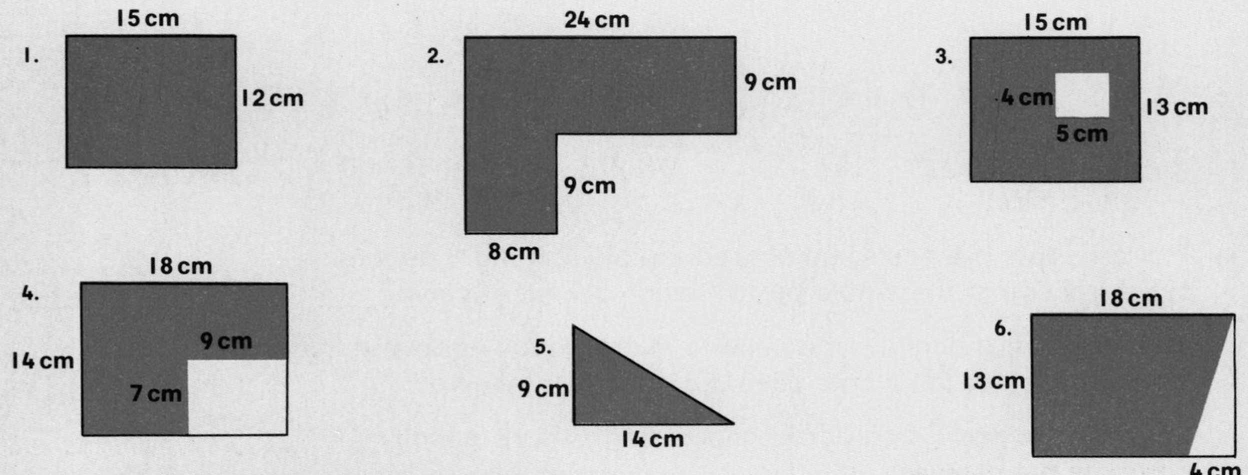

7. Calculate the area of a 13 cm square.
8. Calculate the length of a rectangle whose area is 96 cm^2 and whose width is 8 cm.
9. Calculate the perimeter of a square whose area is 81 cm^2.
10. A rectangle has an area of 48 cm^2. If the length of each side is doubled, calculate the new area.

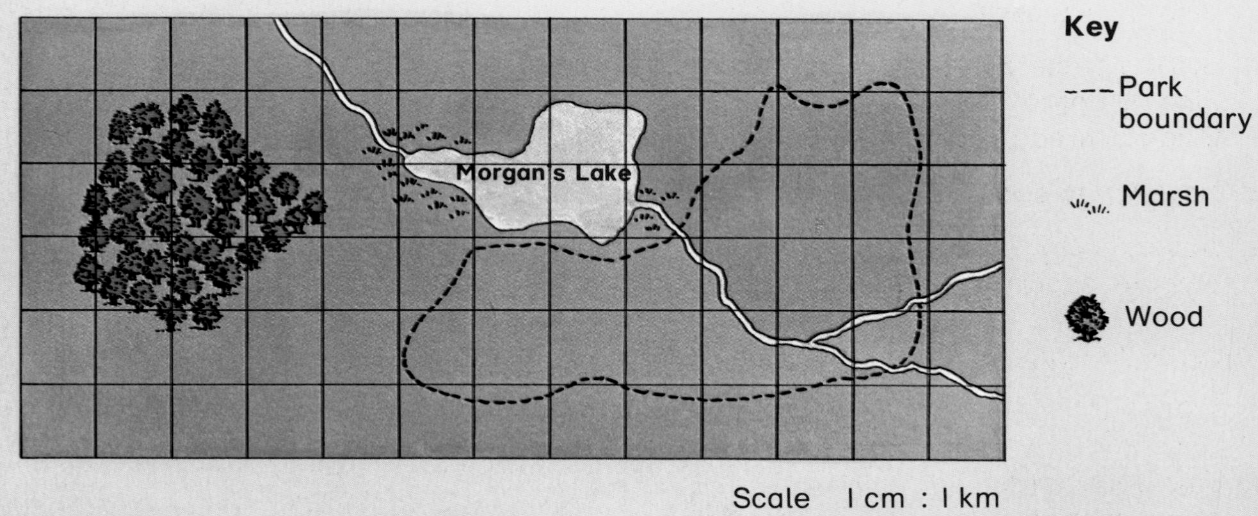

Find the approximate area in km^2 of:

11. the wood;
12. Morgan's Lake;
13. the park.

Cylinder, cm² paper

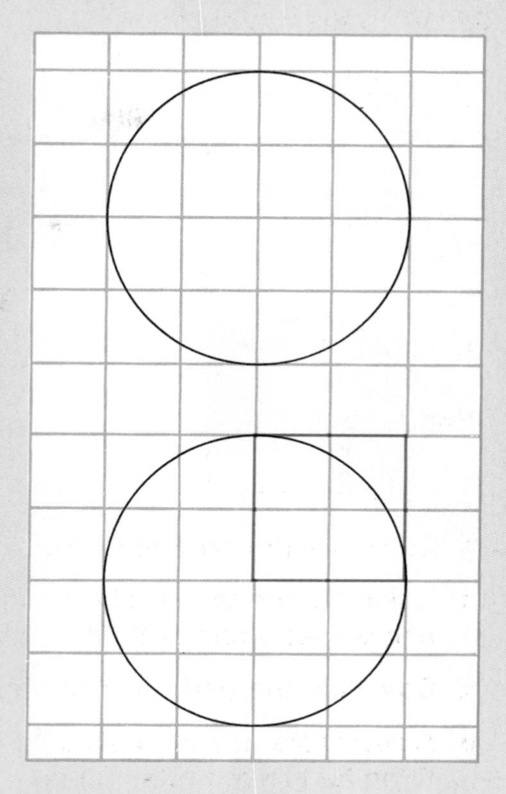

By counting squares find the approximate area of this circle.

1. ≫ Area of circle ≃ ☐ cm²

A square has been drawn on a radius.

2. ≫ Area of square on radius = ☐ cm²

3. Approximately how many times larger is the area of the circle than the square?

4. Draw a circle of radius 5 cm.
   Find its approximate area by counting.
   Find the area of the square on a radius.

5. Now do the same for circles of radius 4 cm and 6 cm.

6. Approximately how many times greater is the area of each circle than the square on its radius?

Calculate the approximate areas of circles with these radii.

7. 7 cm. 8. 10 cm. 9. 15 cm. 10. 23 cm.

11. Find a cylinder.
    Measure its radius.
    Calculate the approximate area of its base.

Percentages

Find:

1. 30% of £50
2. 80% of £10
3. 70% of £20
4. 60% of £5
5. 40% of £25
6. 20% of £10
7. 90% of £50
8. 40% of £10
9. 75% of £400
10. 50% of £300
11. 80% of £250
12. 70% of £500

Find 7% of £15.

If the percentage cannot be changed into a simple fraction, change it into hundredths and multiply.

$$\frac{15}{1} \times \frac{7}{100}$$

$$= \frac{105}{100}$$

$$= 1{\cdot}05$$

7% of £15 is £1·05.

13. 6% of £48
14. 18% of £45
15. 8% of £55
16. 32% of £45
17. 12% of £62
18. 4% of £95
19. 28% of £34
20. 43% of £12
21. 65% of £6

Morton Building Society
gives 12% interest!
Deposit your money here.

What interest would you receive from the Morton Building Society if you deposited these amounts for one year?

22. £5
23. £12
24. £20
25. £36

Bixton Building Society
gives 15% interest!
None can beat us.

What interest would you receive from the Bixton Building Society if you deposited these amounts for one year?

26. £8
27. £14
28. £24
29. £30

3 out of 20 can be written as $\frac{3}{20} = 15\%$
We can work it out this way

$$\frac{3}{\cancel{20}_1} \times \frac{\cancel{100}^5}{1}$$

$$= 15\%$$

To change a fraction into a % multiply by 100.

Write these as percentages.

1. 24 out of 40
2. 3 out of 15
3. 7 out of 20
4. 42 out of 70
5. 14 out of 35
6. 6 out of 24
7. 12 out of 60
8. 8 out of 80
9. 14 out of 28
10. 7 out of 35
11. 9 out of 45
12. 8 out of 40

Here are the end of term marks that Richard got.

Mathematics	44 out of 50
English	18 out of 25
History	14 out of 20
Geography	21 out of 30
Science	7 out of 28

13. Write as a percentage Richard's marks for each subject.
14. At which subject did Richard do best?
15. Richard had the same percentage in two subjects.
    Name the two subjects.
16. What was Richard's average percentage for the five subjects?

Graphs

This graph shows two sets of information.
The red line shows the distance travelled by a lorry.
The blue line shows the distance travelled by a car.

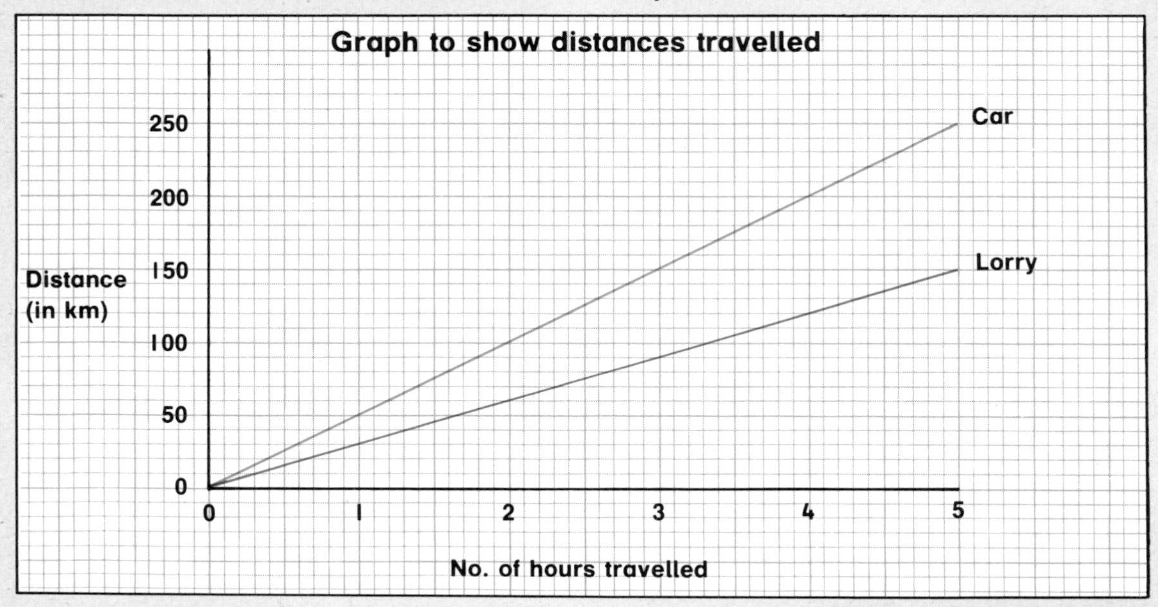

Find this information from the graph.

How far had the lorry travelled after

1. 3 hours?
2. 4 hours?
3. $4\frac{1}{2}$ hours?

How far had the car travelled after

4. 5 hours?
5. $2\frac{1}{2}$ hours?
6. $4\frac{1}{2}$ hours?

How far were the vehicles apart after

7. 5 hours?
8. 3 hours?
9. $1\frac{1}{2}$ hours?
10. $3\frac{1}{2}$ hours?

Diesel costs 40p a litre.
Petrol costs 35p a litre.

Copy these axes.

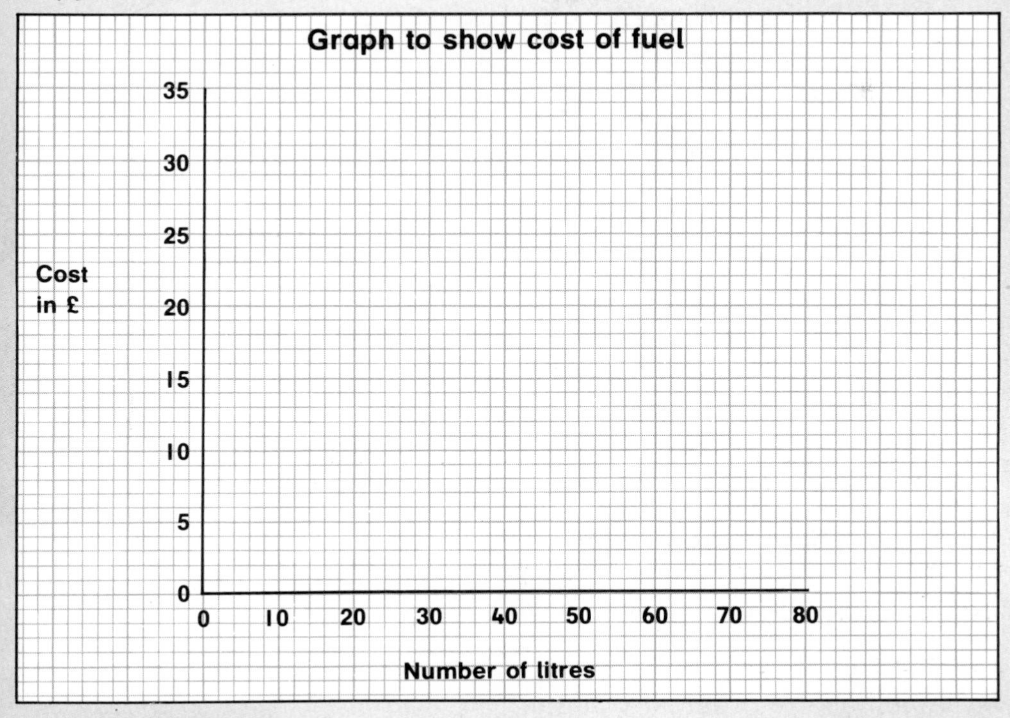

On the axes draw:

1. a graph to show the cost of diesel up to 80 litres.
2. a graph to show the cost of petrol up to 80 litres.

From the graph find the difference between the cost of petrol and diesel for:

3. 20 litres.
4. 60 litres.
5. 10 litres.
6. 50 litres.
7. 30 litres.

Volume

Remember: volume of a cuboid = area of base × height

Calculate the volume of these cuboids.
All measurements are in cm.

1.

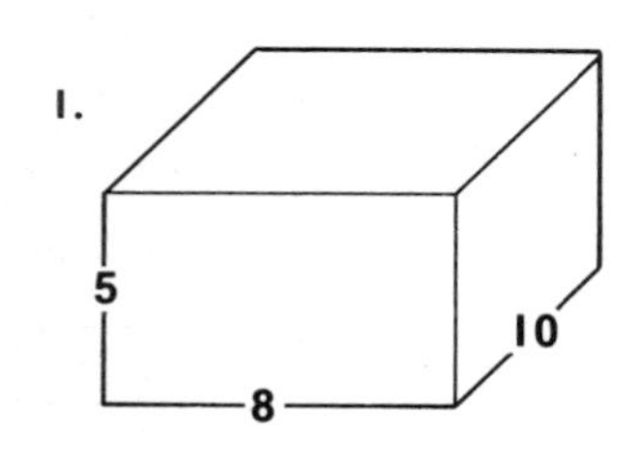

2.

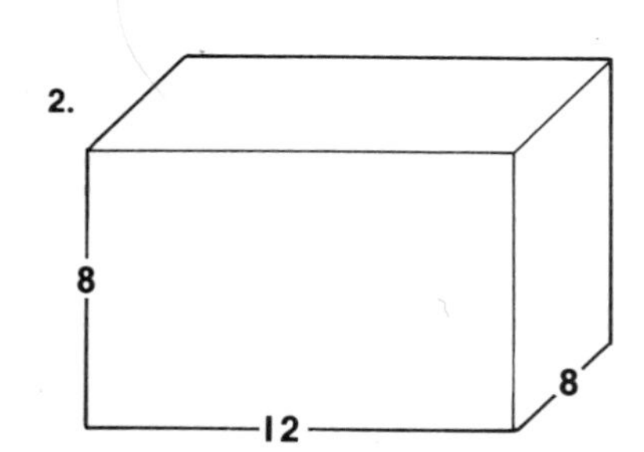

3.

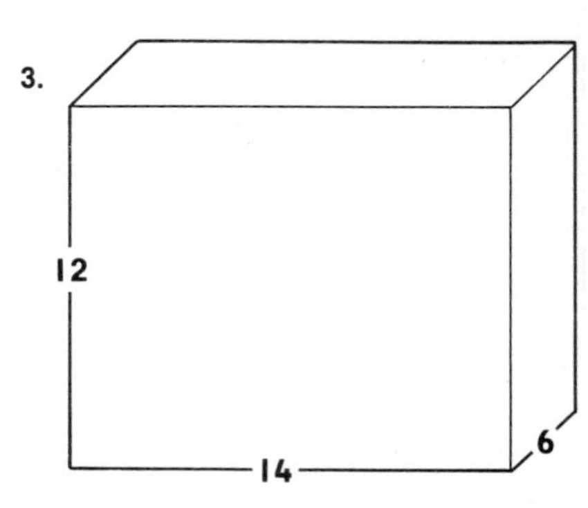

4.

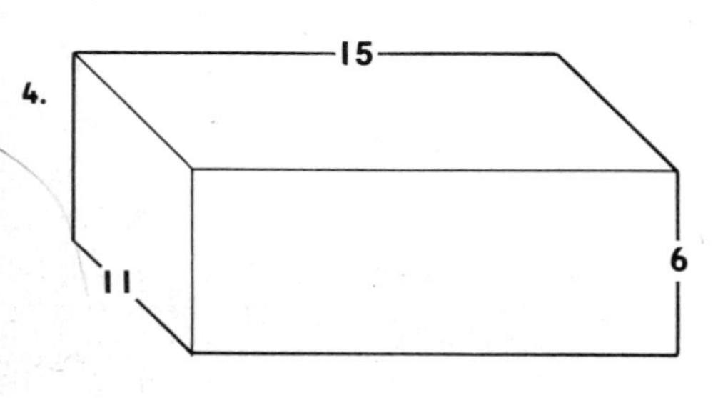

5.

6. Which box do you think has the greatest volume?

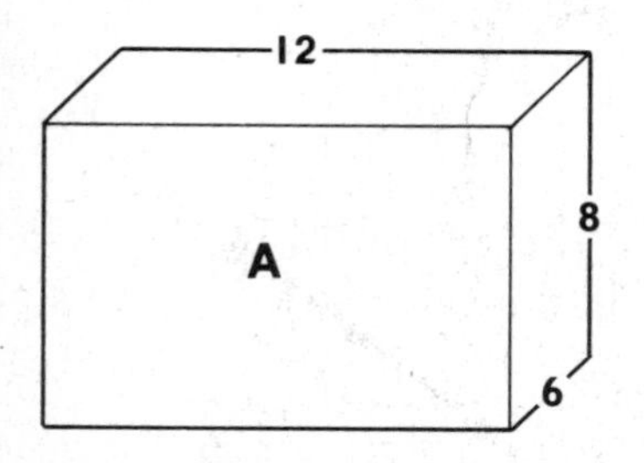

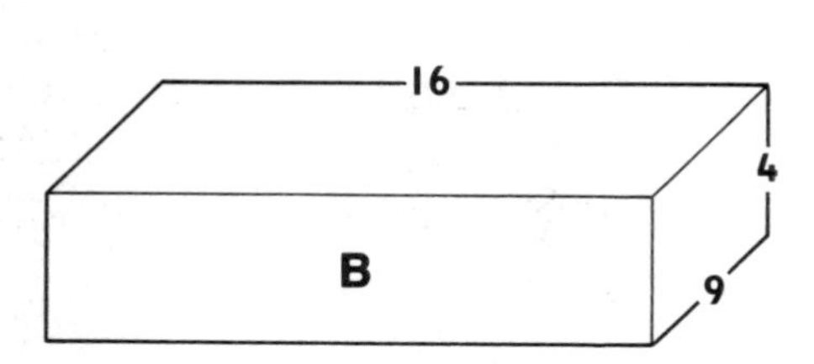

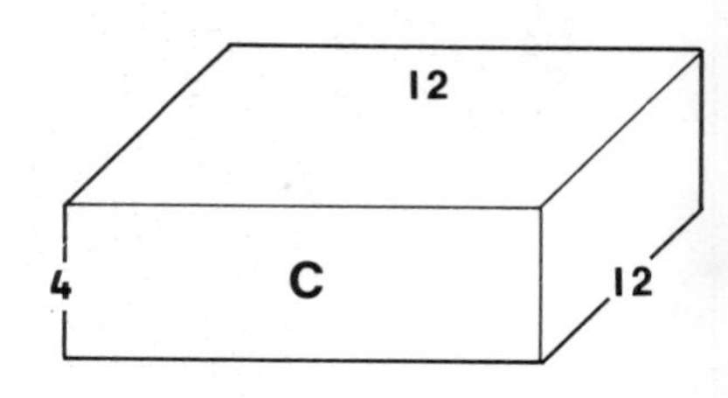

Calculate the volumes to see if you were correct.

7. Calculate the volume of a 12 cm cube.

8. A cuboid has a volume of 512 cm^3.
Its height and length are both 8 cm.
Calculate its width.

9. The area of the base of a cuboid is 24 cm^2.
The volume of the cuboid is 312 cm^3.
Calculate the height of the cuboid.

Hollow 10 cm cube, measuring cylinder

1. What is the volume of the 10 cm cube in cm^3?
2. What is the capacity of the 10 cm cube in ml?

Tell your teacher what you notice.

Find the volume of several containers in cm^3 using a measuring cylinder.

Part fill a measuring cylinder.
Record the water level in the cylinder.
Place a stone completely in the water.
Record the new water level.
The difference between the water levels is the volume of the stone in cm^3.
Measure the volume of other shapes in the same way.

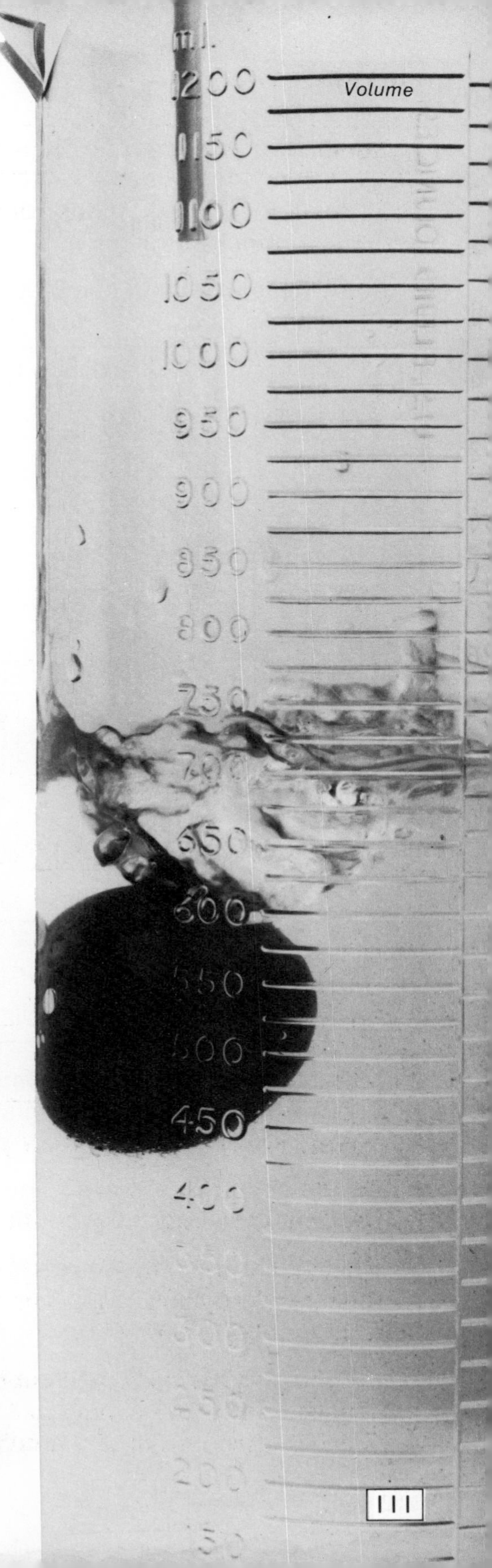

Graphs

Mr. Cope is a bricklayer.
He built a new wall one morning.
He started at 7 o'clock and finished at 1 pm.
He laid 100 bricks an hour.

1. Copy these axes and complete the line graph to show the number of bricks he laid.

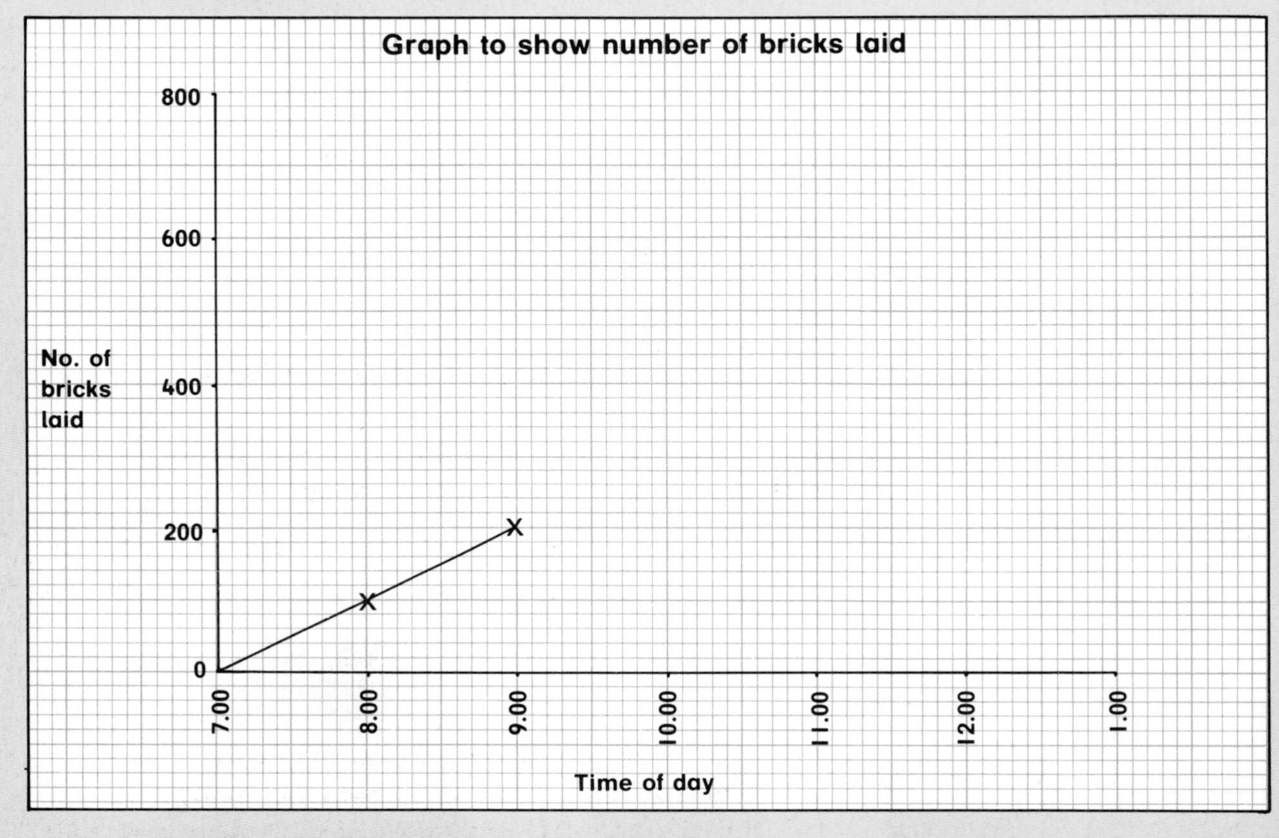

Mr. Bond built a wall on the same morning.
He started work at 8 am and finished at 1 pm.
He laid 125 bricks an hour.

2. On the same axes draw a line graph to show the number of bricks he laid. Remember he started work at 8 am.

3. At what time during the morning had both men laid the same number of bricks?

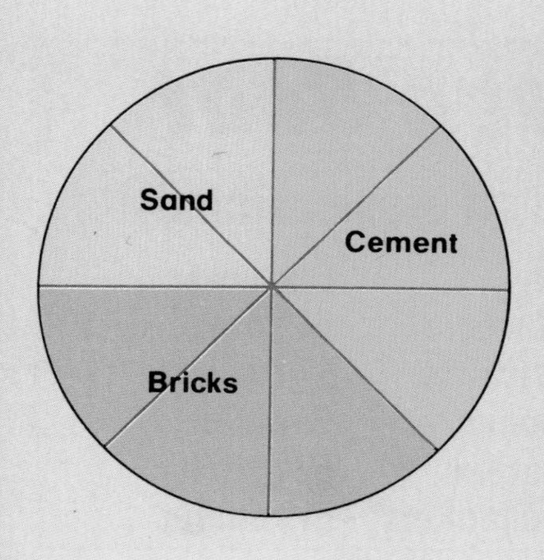

This circle is divided into 8 equal parts.
It is a diagram to show information.
It is called a **pie chart**.
Pie charts are always drawn as circles.

The materials for Mr. Bond's wall cost £96.
This chart shows how his money was spent.
It shows that $\frac{3}{8}$ of it was spent on cement.

1. What fraction of his money was spent on bricks?
2. What fraction of his money was spent on sand?
3. How much did he spend on each item?

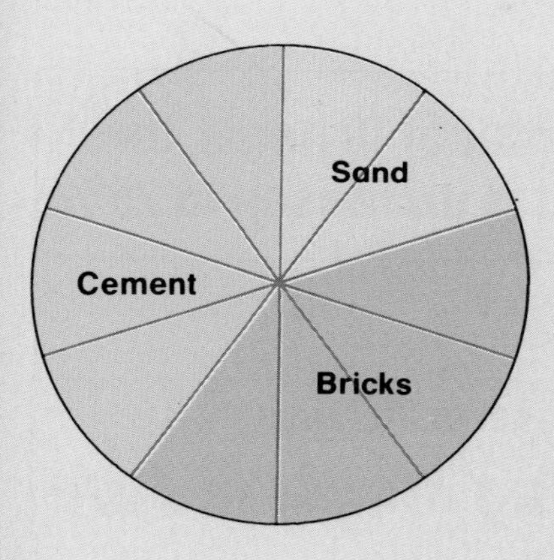

This pie chart is divided into 10 equal parts.

The materials for Mr. Cope's wall cost £120.
This pie chart shows how his money was spent.

4. How much did he spend on each item?

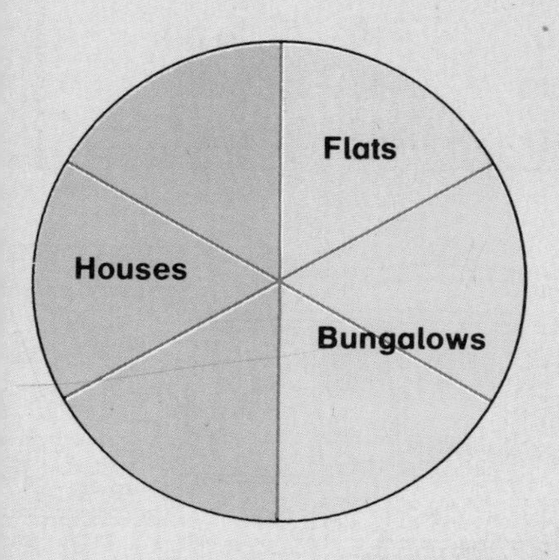

Mr. Bond and Mr. Cope also build homes.
In a year they built 120 homes between them.
This pie chart shows how many of each type they built.

5. From the chart, complete this table:

Type of home	Houses	Flats	Bungalows
Number built			

Percentages

1. Find the cost of 2 tee shirts.
2. What is the cost of a pair of jeans and a tee shirt?
3. How much would it cost for dungarees and two pairs of pop socks?
4. How much more do dungarees cost than a denim jacket?
5. What change would you have from £10 if you bought a tee shirt and a pair of pop socks?

Teen Beat

20% off these prices!!!

Jeans	£19 a pair
Tee shirts	£4 each
Sweatshirts	£7 each
Pop socks	£3 a pair
Dungarees	£24 each
Denim jackets	£19 each

6. What would be the cost of 10 sweatshirts?
7. What would be the total cost of 10 pairs of pop socks and 10 tee shirts?

For an order of 10 items, the discount is increased to 25%

8. Calculate the sale price of each washing machine.
9. Which washing machine is the best bargain?

Graphs

The children at Chartwell School had their photographs taken.
Each photograph cost £4.
This graph shows how much money was collected in each class.

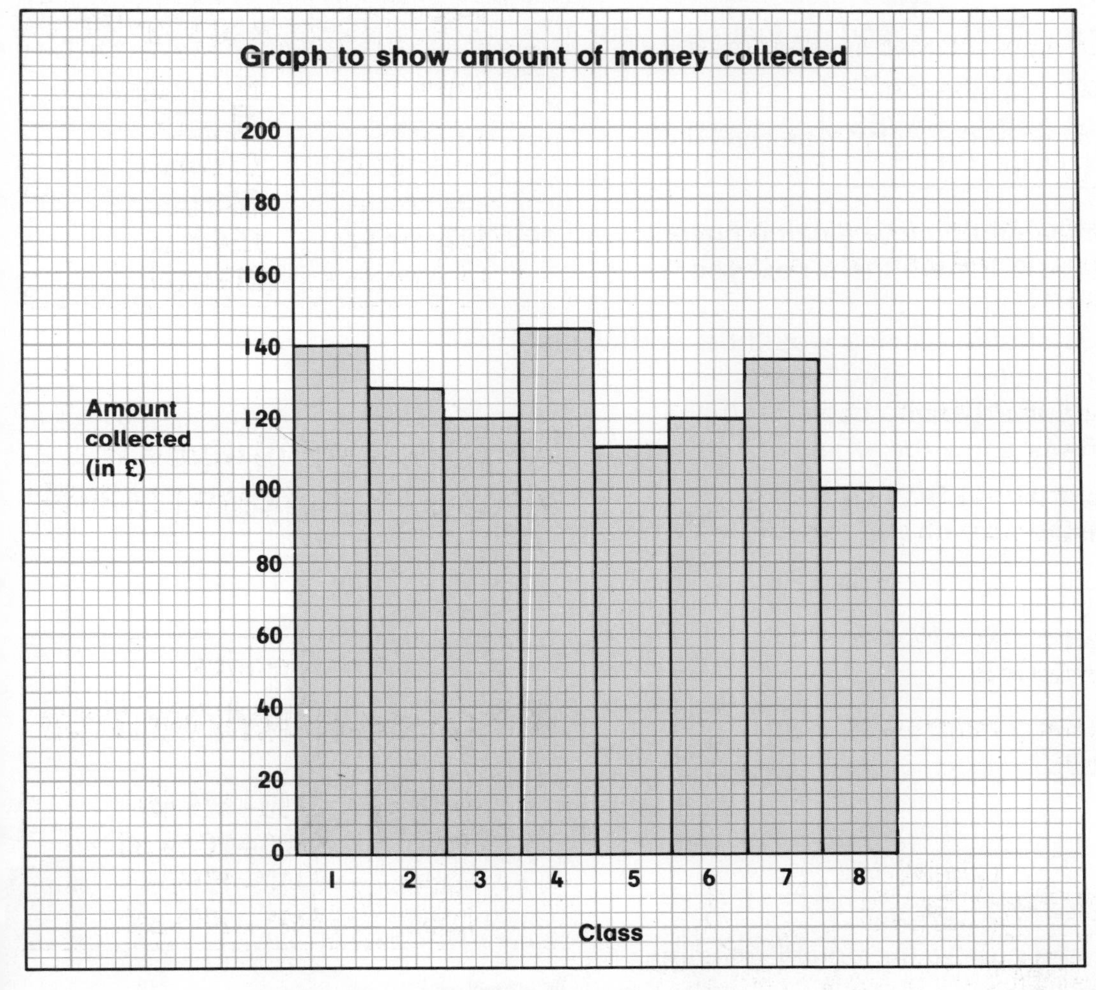

1. How much money was collected in each class?
2. How much money was collected altogether?
3. How much more money was collected in Class 4 than in Class 5?
4. How much more money was collected in Class 3 than in Class 8?
5. What was the average amount collected for each class?
6. How many classes collected more than the average amount?
7. How many classes collected less than the average amount?
8. Write the amount collected in each class as a percentage of the total amount.

Area

Mr. and Mrs. Fleetwood live in a bungalow.
They spent £500 on fuel bills last year.
The cost of fuel is due to increase by 12% next year.

1. How much will they spend on fuel next year?

DOUBLE GLAZING SAVES 5% ON FUEL BILLS

2. How much would double glazing have saved them last year?
3. How much will it save them next year?

They decided double glazing would be worthwhile.
Its cost is calculated on area.
Each m² of double glazing costs £150.
These diagrams show the windows to be double glazed.

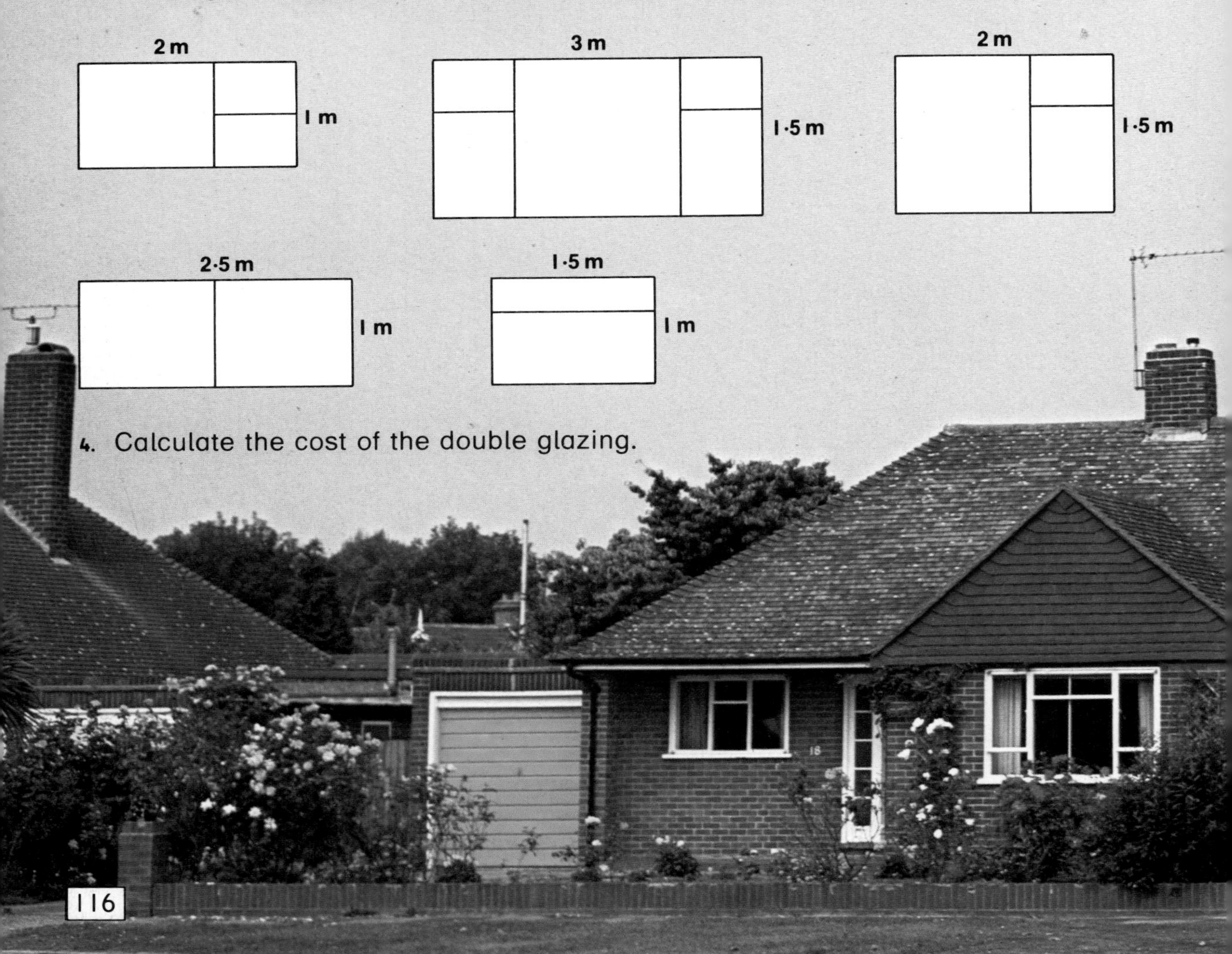

4. Calculate the cost of the double glazing.

Mr. and Mrs. Fleetwood bought carpets for their bungalow.

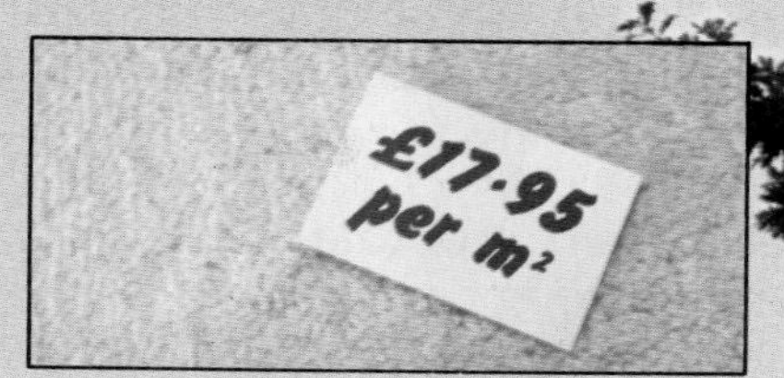

Here is a plan of the bungalow. Scale 1 cm : 1 m.

Kitchen | Bedroom | Lounge | Bathroom

1. What is the area of the lounge?
2. They bought the dearest carpet for the lounge. What did it cost?
3. What is the area of the bedroom?
4. They bought the cheapest carpet for the bedroom. What did it cost?

Measurement

You will need a friend to help you.

Find your "Index of Elegance".
Do it in this way:

Measure the total length of your arm and hand.
Measure the length of your middle finger.
Divide your finger length into your arm length.
The answer is your index of elegance.
Who has the greater index of elegance, you or your friend?

Find your "Index of Strength".
Do it in this way:

Measure round your biceps.
Measure round your chest.
Divide your biceps' measurement into your chest measurement.
The answer is your index of strength.
Who has the greater index of strength, you or your friend?

Find your "Index of Power".
Do it in this way:

Measure round your calf.
Measure how far you can jump from a standing start.
Divide your calf measurement into your jump length.
The answer is your index of power.
Who has the greater index of power, you or your friend?

Number

1. 650 + 280
   If I add a thousand to each number, my answer will be

   a) 1000 more. b) 2000 more. c) the same.

2. 2580 − 1629
   If I add a thousand to each number, my answer will be

   a) 1000 more. b) 1000 less. c) the same.

3. 274 × 10
   If I double each number, my answer will be

   a) twice as big. b) four times as big. c) half as big.

4. 4212 ÷ 4
   If I halve each number my answer will be

   a) half as big. b) twice as big. c) the same.

Write the factors of each of these numbers.

5. 20 6. 12 7. 16 8. 25 9. 27 10. 24

11. 18 12. 15 13. 21 14. 30 15. 14 16. 40

Some numbers have no factors other than themselves and 1.

The only factors of 11 are 11 and 1.
The only factors of 13 are 13 and 1.

Numbers which have no factors other than themselves and 1 are called **prime numbers**.

17. Make a list of all prime numbers under 20.
    (1 is not usually called a prime number.)

18. Write which of these are prime numbers.

    28, 16, 5, 23, 19, 18, 21, 27, 11, 29, 32.

19. Explain why each of these sentences is true.

    An even number greater than 3 can never be a prime number.
    A square number can never be a prime number.
    A number ending in 5 can never be a prime number.

Measurement

Balance, weights in pounds and ounces

Imperial units

When your parents were at school they had to weigh objects using pounds and ounces.

Ounces

A short way of writing **ounce** is **oz**.
4 ounces = 4 oz.

Find the weights marked 1 oz, 2 oz, 4 oz and 8 oz.
Using a balance and ounces find the weights of four objects.
Record your results.

Pounds

A short way of writing **pound** is **lb**.
2 pounds = 2 lb.

Find the weights marked 1 lb, 2 lb and the ounce weights.
Using your balance find how many ounces equal one pound.

≫ ☐ ozs = 1 lb.

Now find four objects to weigh using pounds and ounces.
Record your results.

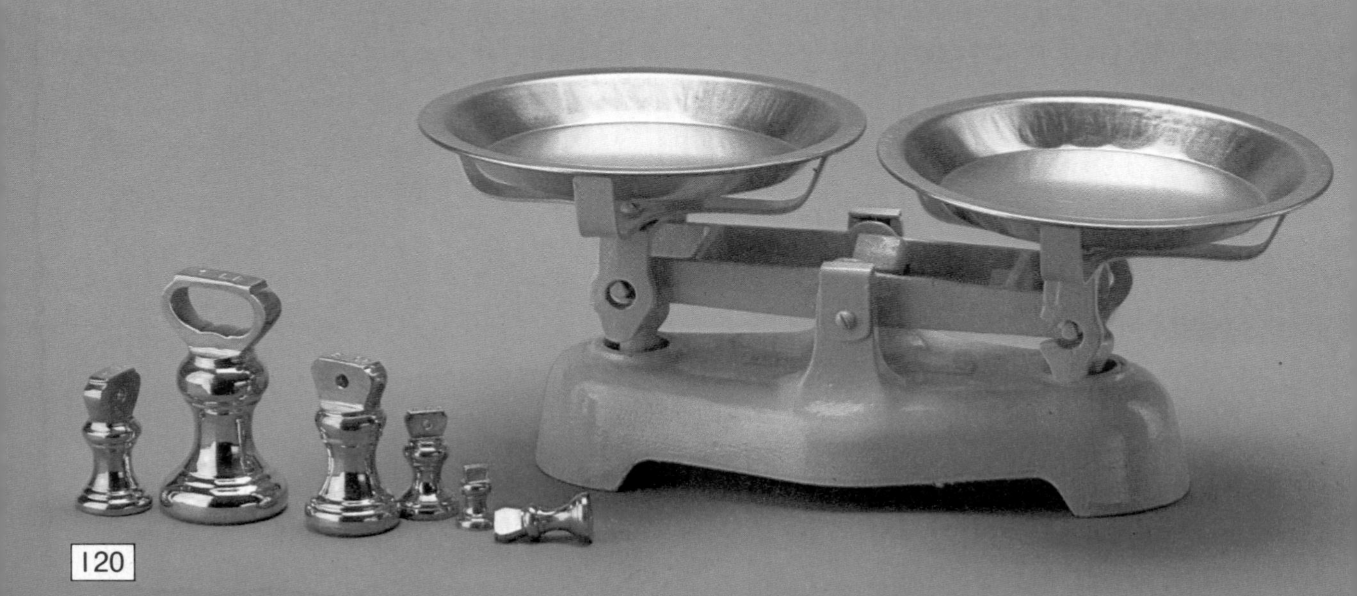

Highfield Hall School has a swimming pool for learners.
The bottom of the pool does not slope.
It measures 16 m long, 5 m wide and 1 m deep.

1. What is the volume of the pool?

2. The pool is $\frac{3}{4}$ filled with water.
   What volume of water is in the pool?

3. 1 m^3 of water = 1000 l.
   How many litres of water are used?

4. 1 l of water = 1 kg.
   What weight of water is in the pool?

5. 1000 kg = 1 tonne.
   Find the weight of water in the pool in tonnes.

6. Chemicals are added to the water.
   500 ml of chemicals are added to each 10000 l.
   How many ml of chemicals have to be added to the pool?

7. A swimming lesson lasts for $\frac{3}{4}$ hour.
   The children take $\frac{1}{9}$ of the lesson to undress and $\frac{2}{9}$ of the lesson to dress.
   How many minutes are spent swimming?

8. For each lesson 24 children are allowed in the pool.
   The school has 288 pupils. How many swimming lessons are needed for each child to swim once a week?

9. What is the total time spent on swimming lessons during a week?

Investigations

Put a counter at the beginning of the maze.
At each junction toss a coin.
If the coin shows heads, move down and to the left.
If the coin shows tails, move down and to the right.
Where did your counter leave the maze? At A, B, C or D?

If you repeat this activity 25 times, which of the four exits do you think will be used most?
Work your way through the maze 25 times.
Record the exit used each time.
Were you correct?

Which is the most popular letter of the alphabet?

Choose any reading or reference book.
Choose any ten complete lines of words and count how often each letter of the alphabet occurs.
Keep a tally of your findings.

Which letters are used most often?
Which letters are used least often?

Letter	Tally	Total
A	IIII	
B	I	
C	III	

An American inventor, Samuel Finley Breese Morse, who was born in 1791, invented a simple dot and dash code for sending messages by telegraphy. Here is his code.

A .—
B —...
C —.—.
D —..
E .
F ..—.
G ——.
H ....
I ..
J .———
K —.—
L .—..
M ——
N —.
O ———
P .——.
Q ——.—
R .—.
S ...
T —
U ..—
V ...—
W .——
X —..—
Y —.——
Z ——..

Look at your results of the most popular letters.
Look at the Morse Code for those popular letters.
What do you notice?

Look at your results of the least popular letters.
Look at the Morse Code for these letters.
What do you notice?

Calculator

Bull's Eye
A game for two players and one calculator.

Rules

1. Each player secretly writes any number he wishes as his target score.
2. Each player has the four signs to use (+ − × ÷), and can only use any one of them once in the game.
3. Any of the nine digits may be used only once during the game.
4. After four turns each the game is over.
5. The player whose target score is closest to the final number on the calculator wins the game.

Here is a game being played by Helen and Craig.

Helen chooses 26 as her target score.
She has + − × ÷ to use.

Craig chooses 3 as his target score.
He has + − × ÷ to use.

Turn 1 Digits to be used: 9 8 7 6 5 4 3 2 1

She chooses 8.
(So 8 cannot be used by either player again.)

He chooses × 2.
(2 cannot be used again and he has used his × sign.)

Turn 2 Digits left to be used: 9 7 6 5 4 3 1

She chooses + 9.
(9 has now been used up so has her + sign.)

He chooses ÷ 7.
(7 has been used up so has his ÷ sign.)

Turn 3 Digits left to be used: 6 5 4 3 1

She chooses × 5.

He chooses + 1.

Turn 4 Digits left to be used: 6 4 3

She chooses − 3.

He chooses − 4.

Who won? You play the game.

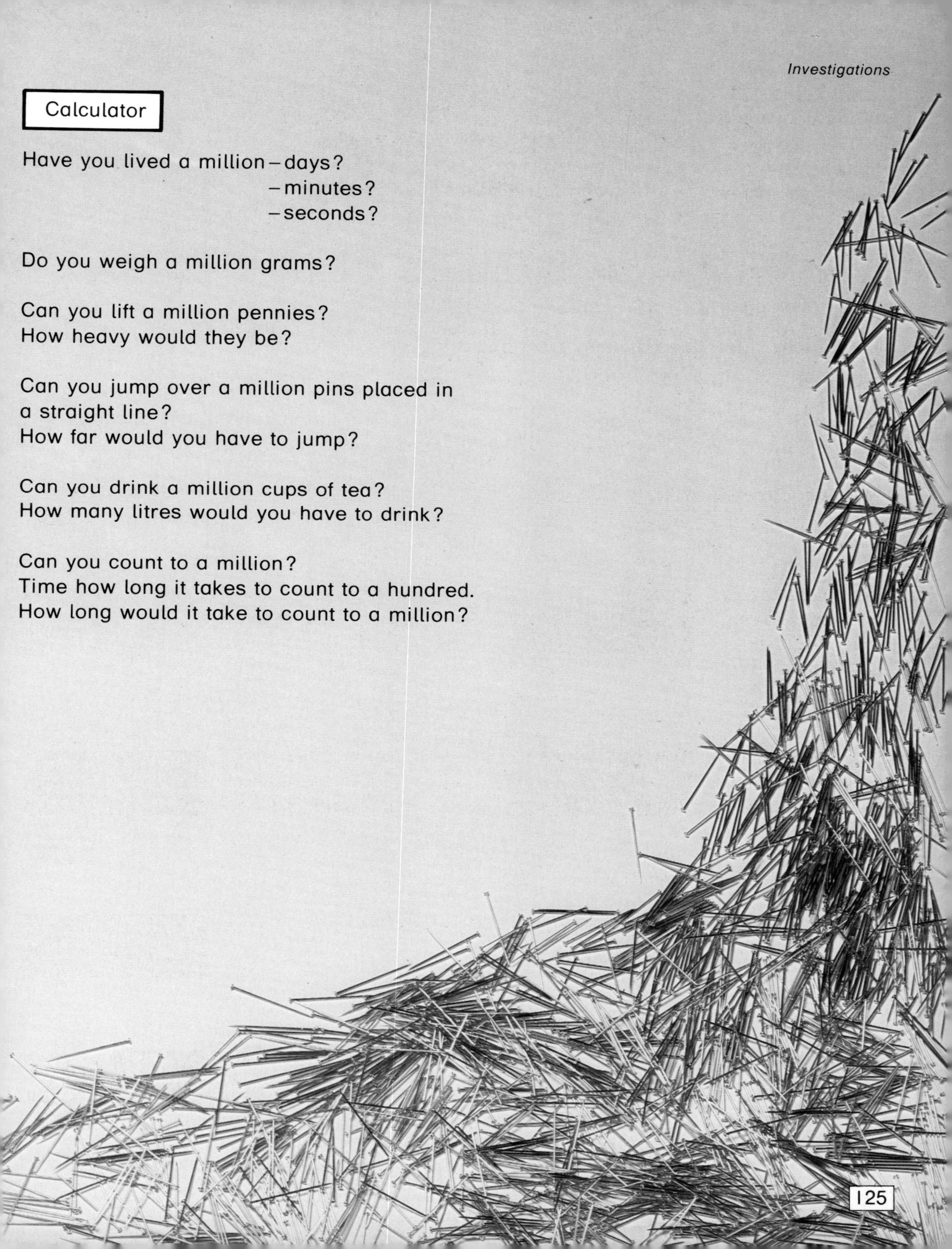

Calculator

Have you lived a million – days?
– minutes?
– seconds?

Do you weigh a million grams?

Can you lift a million pennies?
How heavy would they be?

Can you jump over a million pins placed in a straight line?
How far would you have to jump?

Can you drink a million cups of tea?
How many litres would you have to drink?

Can you count to a million?
Time how long it takes to count to a hundred.
How long would it take to count to a million?

Assessment

1. $\begin{array}{r} 4789 \\ 1298 \\ +\ 4706 \\ \hline \\ \hline \end{array}$

2. $\begin{array}{r} 6240 \\ -\ 1397 \\ \hline \\ \hline \end{array}$

3. $\begin{array}{r} 427 \\ \times\ \ 36 \\ \hline \\ \hline \end{array}$

4. $9\overline{)4833}$

5. Add £7·53, £2·26 and £0·85.
6. Find the total of 7·5 kg, $6\frac{1}{4}$ kg and 3·452 kg.
7. Find the difference between 7·5 m and $3\frac{3}{4}$ m.
8. 1·400 l × 8.
9. Find the area of this shape.

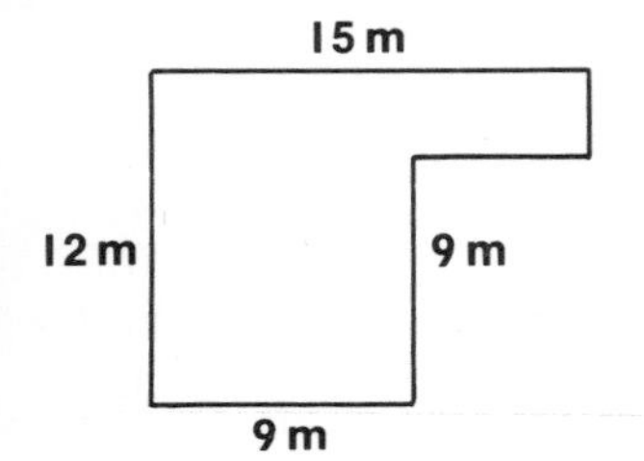

10. Find the volume of this cuboid.

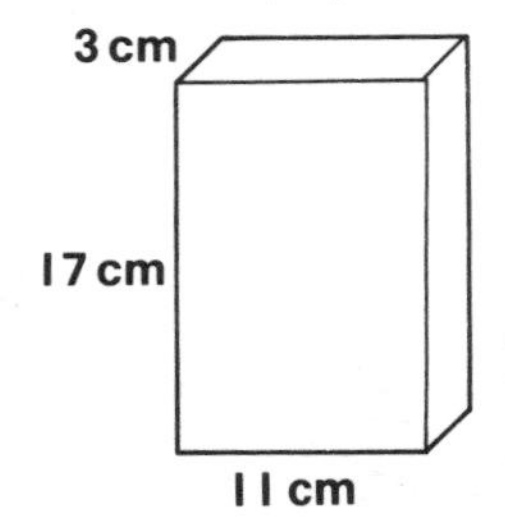

11. Write these fractions in order of size.

$\frac{7}{12}, \frac{2}{3}, \frac{3}{4}$

12. $2\frac{9}{10} + 1\frac{3}{4} + 4\frac{1}{2}$
13. $3\frac{1}{6} - 1\frac{3}{4}$
14. $1\frac{2}{7} \times 4\frac{2}{3}$
15. $\frac{2}{7} \div 6$
16. $54 \times \frac{1}{9}$
17. Write $\frac{3}{4}$ as a decimal.
18. Divide 2·76 by 4.
19. Find 12% of £45.
20. Write 11 out of 55 as a percentage.
21. Which of these shapes are prisms?

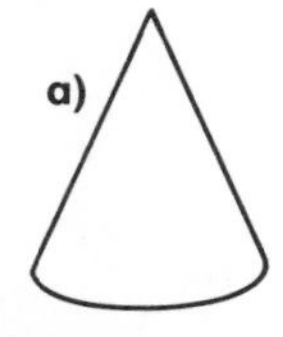

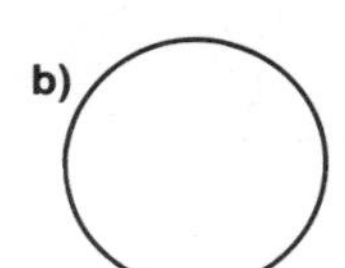

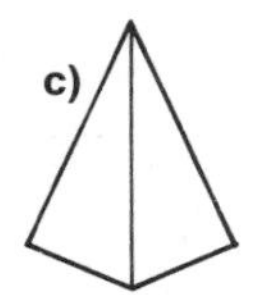

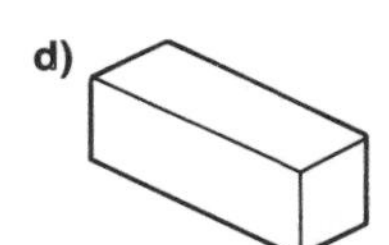

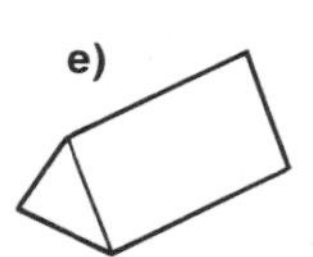

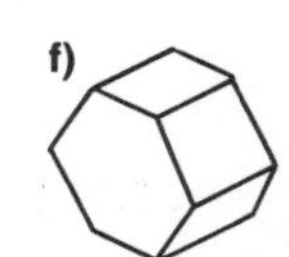

22. Which of these are prime numbers?

14 28 3 7 18 11 20

23. $34\overline{)8806}$

24. Write the value of the underlined digit 7<u>6</u>217.

25.

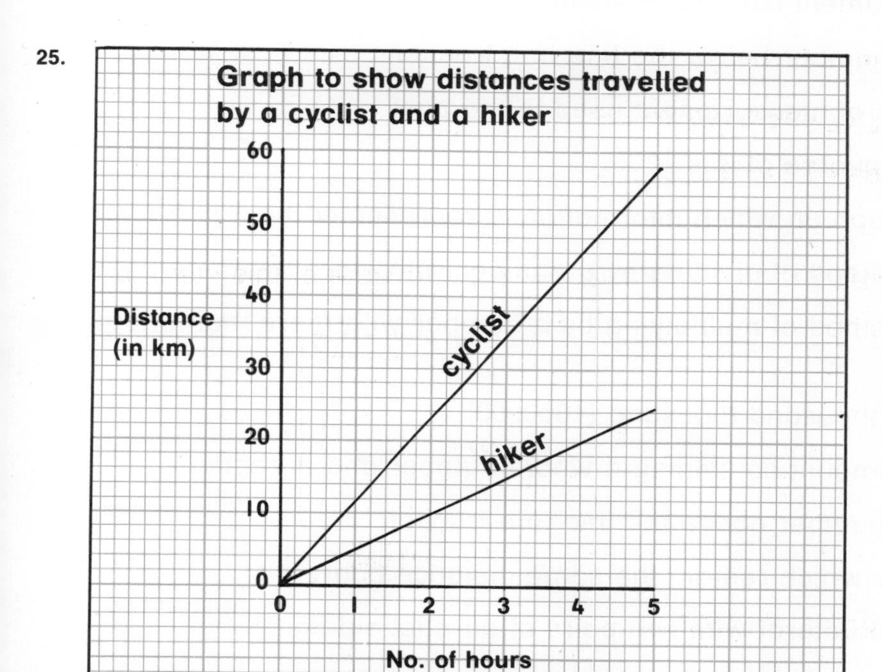

How far apart were they after 4 hours?

26. How many minutes from 1140 to 1315?

27. Map the quadrilaterals to their names.

parallelogram

trapezium

rhombus

rectangle

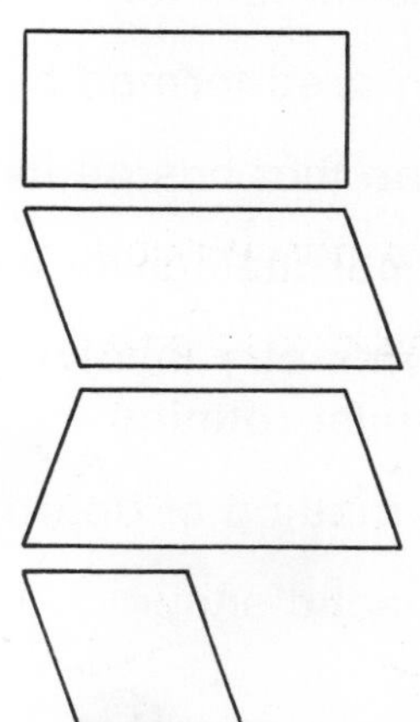

Glossary

angle of elevation	angle measured from the horizontal
calculator	instrument used for calculations
cardioid	pattern made up of circles
clinometer	instrument used for measuring angles of elevation
denominator	the number below the line in a fraction
Imperial units	units of measurement used in the past
inverse	the inverse of 8 is $\frac{1}{8}$
line graph	a graph on which information is represented by a line
long division	a method of dividing by a number with more than one digit
long multiplication	a method of multiplying by a number with more than one digit
million	one thousand thousand (1000000)
mixed number	an amount containing a whole number and a fraction
numerator	the number above the line in a fraction
parallel lines	lines which remain the same distance apart
parallelogram	quadrilateral with two pairs of parallel sides
pendulum	consists of a weight hung so that it can swing
per cent (%)	out of a hundred ($10\% = \frac{10}{100}$)
pie chart	a circular diagram to show information
prism	a solid which is the same shape all the way through
protractor	instrument for measuring angles
region	an area formed by lines intersecting
regular shape	one that has all its sides equal and all its angles equal
rhombus	a parallelogram with four equal sides
rotational symmetry	symmetry involving a shape fitting into its own outline whilst rotating
rounding off	a method of approximating
tetrahedron	a solid shape